TRADING THE GENOME

TRADING THE GENOME

Investigating the
Commodification of
Bio-Information

BRONWYN PARRY

COLUMBIA UNIVERSITY PRESS   NEW YORK

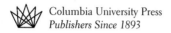

Columbia University Press
Publishers Since 1893

New York   Chichester, West Sussex
Copyright © 2004 Columbia University Press
All rights reserved

Library of Congress Cataloging-in-Publication Data
Parry, Bronwyn.
    Trading the genome: investigating the commodification of bio-information /
Bronwyn Parry.
        p.   cm.
    Revised ed. of author's PHD thesis entitled The fate of the collections (Univ.
of Cambridge, 1998).
    Includes bibliographical references and index.
    ISBN 0–231–12174–1 (cloth : alk. paper)
    1. Biotechnology—Social aspects. 2. Germplasm resources.
3. Bioinformatics. I. Title.

    TP248.23.P376 2004
    303.48'3—dc22
                                                                        2004045456

Columbia University Press books are printed on
permanent and durable acid-free paper.
Printed in the United States of America

c 10 9 8 7 6 5 4 3 2 1

For Jack, Win, and Juliet

CONTENTS

ILLUSTRATIONS

TABLES

ACKNOWLEDGMENTS

This book, like its subject matter, has been a product of constant reimagining and re-engineering. It began life as a doctoral dissertation at the University of Cambridge, but has since been reworked and refined to take account of phenomena that are, themselves, subject to constant and highly complex processes of change. It would not have been possible for me to produce this work without the continued support of a number of individuals and institutions. Professor Richard Howitt first set me on the path to Cambridge by instilling in me a deep respect for work that is both intellectually rigorous and imaginative and for helping to expand my (then) limited conception of what it is to think "geographically." Professor Linda McDowell provided me with all the strengths of a Cambridge education, encouraging me to think "outside of the box" while simultaneously tempering my more fanciful tendencies with her incisive critique and editorial skill. Beyond this, I am also indebted to her for introducing me to Newnham College, Cambridge. Traveling halfway across the world to undertake a Ph.D. might have been a very alienating experience had I not been made to feel so welcome and supported there by friends such as Eldrid Herrington, Andrew Taylor, Karen McAuley, Kate Dailinger, Cate Hancock, Julia Hoggett and Clare Drury.

More recently, I have benefited enormously from holding a research fellowship at King's College, Cambridge. I am particularly indebted to King's, firstly, for recognizing the importance of undertaking research of this nature and, second, for providing me with the funding and intellectual and material space in which to conduct it. Their continued support has enabled me to convert the dissertation into a book and to extend my research into new domains. It is hard to imagine how I could ever expect to find a more supportive or intellectually challenging work environment.

This research was interdisciplinary in nature, and, as such, it required me to acquaint myself with a number of unfamiliar literatures and epistemologies. I could not have managed this without the patient tutelage of a number of colleagues and friends whose specialties lie in different disciplines: in philosophy and the history and philosophy of science, I owe particular thanks to Cathy Gere, Onora O'Neill, Jim Secord, Neil Manson, and Peter Lipton; in anthropology: Caroline Humphrey, Cori Hayden, Marilyn

Strathern, and Sarah Franklin; in English: the late Tony Tanner, Maud Ell-
mann, and Iain Patterson; and in molecular biology: Matt Garner. In law, I
would particularly like to thank a triumvirate of Australian women lawyers
at Cambridge: Natasha Cica, Renee Leon, and Susan Marks. In geography,
I would like to thank David Pinder, Phil Howell, Michael Bravo, Laura
Cameron, Kristian Teleki, Gerry Kearns, and Catherine Nash, all of whom
spoke my language and kept me sane throughout. For their help in concep-
tualizing this work, I am most grateful. That said, responsibility for the final
analysis (and any inaccuracies or misconceptions that it might contain) remains mine alone.

I discovered during the course of this research just how much it costs to
do a "multisite ethnography." Investigating global processes really does require travel over great distances, and I could not have undertaken this without the generous financial support of the British Council, The University of
Cambridge, Newnham College, and King's College. I would also like to sincerely thank my friends and colleagues in New York and Washington, D.C.,
including Nora Conant and her family, Art Herrington, Christine Padoch,
and Miguel Vasquez-Pinedo, who provided me with accommodation, wonderful meals, support, encouragement, and transportation during the course
of my long fieldwork in the United States.

I also owe a very special debt of thanks to all of those people in the United States who agreed to be interviewed or surveyed for this study. The sensitivity of the subject matter is such that all took great risks (both personal
and professional) in agreeing to answer my probing questions. Although it
has not been possible for me to name all of them here, they all gave several
hours, or indeed days of their time, to provide a detailed account of their
work practices and relationships, beliefs, and practical and ethical concerns.
Their openness and honesty was, in part, generated by their desire to be re-
flexive about their own practice. I cannot stress how much I appreciate their
willingness to involve me in that project. I sincerely hope that the findings
of this research can feed back into that process in a constructive way.

Columbia has been as generous, accommodating, and encouraging a
press as any first author could possibly hope for. I would like to thank Holly
Hodder for inspiring me with exciting ideas as to the shape that the book
could take and my editor Robin Smith for successfully guiding the manu-
script through the necessary stages to completion. Michael Haskell's copy-
editing, which I think of as a special kind of orchestration, was extraordinarily sensitive and intelligent, so to my own Robert Russell Bennett, my very
sincere thanks. I would like to thank Dr. Jill Whitlock of the Whipple Mu-

seum of the History of Science, Dor Duncan of the Kew Royal Botanical Garden Library, Miriam Gutierrez-Perez at the Wellcome Trust's History of Medicine Library, Jessica Schwarz of Columbia University Press, and Owen Tucker in the Cartography Office of the Department of Geography at the University of Cambridge for their very kind help in selecting and reproducing the images. My thanks also go to the two anonymous readers whose enthusiasm and feedback on a first draft provided some much appreciated assurance and valuable guidance.

On a more personal note I would like to thank Rosie Harris, Trudy Dalgleish, Sue Fern, Wendy Brown, Jenny Vuletic, Ian Bryson, Trish Barker, and Sam Causer for their continued support. I would also particularly like to thank Flora Gathorne-Hardy. Her intellectual curiosity, belief in the thesis, empathy, and good humor enabled all manner of ideas to be realized here. Finally, I would like to thank my family. My parents, Jack and Win have offered unconditional love and support over many years, as have my wonderful sisters Sue, Ann, and Jill, while the Davis family provides me with a very special home away from home, all of which I greatly appreciate. Last, but most important, I want to thank Juliet for her unfailing loyalty and love, for sharing my passions, and for giving me a future I want to run toward.

PREFACE

It is a peculiar fact of life that profound change rarely occurs cataclysmically. In many cases, the revolutions that transform lived experience begin as minor ones that start unnoticed and go on unremarked. Changes occur subtly, almost imperceptibly, and it takes some while before their cumulative and transformative effects become fully apparent. This has certainly been the case with the information-processing revolution and its effect on the way we experience and use particular commodities. If I think back to when I received my first pay envelope, it strikes me as rather odd to recall that it was, actually, an *envelope*, containing cash and coins. This money was put to use straight away—I used it to pay for my rent, my groceries, and a host of other goods and services. Sometime over the past two decades that practice stopped, for me and for millions like me. Now, I receive my salary electronically, and I pay for everything from train tickets to dry cleaning to car repairs electronically. too. A casual enquiry at any supermarket, garage, or restaurant (in Western countries at least) reveals that cashless transactions like these are now very much the norm. This is not to suggest, however, that hard currency is becoming extinct—the recent introduction of the new two-pound coin in England, the Sacagawea "golden dollar" in the United States, and the creation of Euromoney all confirm that it is not. Nevertheless, it is clear that at some point "money" underwent a transformation, becoming accessible in new ways.

Similar processes have been at work in other domains. Earlier in my academic career, less than ten years ago, undertaking research required some physical stamina. This was not because it entailed strenuous fieldwork but rather because it was necessary to cart back from the library armloads of heavy books and other texts. This too is gradually changing. Much of the information that was once available only in "hard copy"—books, journals, and newspapers—is now obtainable in new digital or electronic forms, as Web pages that I can access instantaneously from any computer terminal, for example. Of course, the books, newspapers, and articles continue to enjoy a hard-copy existence in various locations; I just sometimes choose not to access them in that form, as it is more convenient, or faster, or more efficient to download the material directly from the Internet. Interestingly, is now possible for me to acquire and circulate the information contained in a book or a newspaper without ever actually having had possession of that book or newspaper. Equally, it has now become possible for me to download music directly from the Net without ever having to purchase a tape or a compact disc.

In each case, the information that is embodied in, or represented by, the material object (the currency, the book, or the compact disc) has, as a consequence of the application of new technologies, become available to me in new forms and through new media. This is not a new phenomenon. Inventions such as the phonograph, the wireless radio, the camera, the photocopier, and other similar technologies have all served to allow information to be embodied and conveyed in new ways. The new informational technologies—microprocessors, satellites, electronic circuitry, and the like—have converged to create another information-processing revolution that is perhaps then not so different in type as it is in degree. It has been possible, for example, to convey "money" by a wireless transaction for more than a hundred years. However, it is only now that the number of transactions that take this form has become so large, and their use so pervasive, that they begin to fundamentally alter understandings of how money is, or can, be constituted, and how it can be accessed and utilized. Global twenty-four-hour-a-day currency trading, for example, could not have eventuated without the development of technologies that enable multiple thousands of electronic financial transactions to occur instantaneously.

New modes of transaction are appearing in many different domains that are only made possible by the development of technologies that enable information-based resources and commodities to be "rendered" in new, more transmissible forms. This rerendering involves a process of "stripping down" these commodities or resources to what might be thought of as their bare informational essentials. So, for example, the information in books or journals, maps, architectural plans, or musical compositions is extracted and conveyed digitally or electronically as web pages, digital images, or MP3 files. Much of the existing body of the work (as we have known it) is divested; the pages of the book and the binding, the notes and coins, the paper plans and plastic discs are dispensed with in order to render the information they contain in new *and highly transmissible* electronic or digital forms.

In the frenetic marketplace of the early twenty-first century, the ability to access information at great speed provides an important competitive advantage. In the worlds of business and finance, at least, an hour-long expedition to the library to gather information that can now be accessed online is a luxury few can afford. Nonetheless, there remain other consumers who con-

sider the experience of visiting the library and handling the books or journals a more enriching one than that involved in accessing the same material online. Over time, separate markets have emerged for the same commodity in variously constituted forms: for online and hard-copy newspapers,
for live music, for music recorded on tape or compact disc, and for music
conveyed in digital form, as an MP3 file, for example. It is the subject of
considerable debate as to whether these markets compete with or complement each other. What is clear, however, is that the ability to present information in new forms is radically altering the existing market dynamics for
trade in particular resources or commodities, creating new, vibrant
economies in all manner of new entities.

The ability to translate some commodities into highly mobile informational forms has enabled producers to increase the speed at which they can
be circulated around the global economy. It also enables them to replicate,
combine, and modify these new forms with much greater ease. While this
clearly has some distinct advantages, it may also create some complex dilemmas. One of the most serious of these is how to prevent the *unauthorized*
use, replication, or modification of information-based products. This particular dilemma has been amply illustrated in the recent legal actions taken
against Napster. Napster is a computer interface that enables people to locate and download music in a digital MP3 format from the Internet. It does
so by connecting them with other users who have created personal libraries
of MP3 files by copying music from compact discs. A recent U.S. report suggested that traffic in these recordings has grown so dense that some 75 percent of computer use in U.S. universities is taken up in the pursuit of MP3
recordings.[1] As most MP3 recordings are illicit copies, their exchange generates no further royalties for the creators of the work.

Unlicensed duplication of recordings has, of course, been commonplace
for some time; however, the ability to translate music into digital forms that
can be conveyed and downloaded instantaneously via the Internet has dramatically increased both the number of unauthorized recordings and consequent loss of revenue for recording artists, producers, and retailers. At the
"Music on the Net" conference held in San Francisco on 20 September
2000, it was reported that record labels are now losing $3.1 billion annually
in music sales to unauthorized methods of digital distribution.[2] New products are also being created by combining or manipulating information, such
as data and images, that have recently been rendered in new digital or electronic forms. However, the status of many of these "derivatives" remains uncertain. Questions of who should have rights to or interests in works such as

databases that are compiled by combining or manipulating digital or electronic renderings of existing images or data remain hotly contested.

All of this raises the question of whether there are any restrictions on the *types* of information that could be rendered in these new ways and through these new mediums? This becomes immensely significant when we think about the parallels that exist between informational technologies and biotechnologies. Biotechnologies could be said to have a similar effect as other informational technologies, in that they enable biological materials to be stripped down, or rendered, in new more artifactual or even purely informational forms: as cryogenically stored tissue samples, as extracted DNA, as cell lines, MRI scans, or sequenced DNA coded onto databases. These new technological artifacts, (like their counterparts—MP3 files, digital images, e-publications, and the like) privilege the informational attributes of the resource in question (in this case, the genetic or biochemical information embedded in biological material) at the expense of other attributes (such as the majority of the body of the organism), which are divested. When rendered in an either partially or wholly decorporealized form, genetic and biochemical material and information (what I term here "bio-information") becomes infinitely more transmissible, replicable, and manipulable. This is advantageous to many consumers (particularly those working in the life sciences industry) who wish to access and act on this information as quickly and efficiently as possible.

New and potentially very lucrative markets in these forms of bio-information are consequently beginning to emerge. As the science journalist Tim Radford noted recently, these advancements are "fuel[ing] new multi-billion pound industries based on the software of life."[3] The crucial question is: Who is securing control of this new market economy and on what terms? The conflicts that have raged around the public or private ownership of information derived from the Human Genome Project, the controversies surrounding the patenting of gene sequences, and the debate over the commodification of body parts and tissues show that the questions of who "owns" genetic and biochemical materials and information and who should profit from their exploitation remain unresolved. We cannot begin to answer these questions, it seems to me, until a great deal more is first known about how this emerging resource economy in bio-information actually functions. It is particularly important to understand how this new economy will operate, as it has the capacity to create not only new dynamics of biological-resource exploitation but, more importantly, new geographies of justice and injustice. I thus set out in 1995 to investigate how "bio-informational" resources are being collected and transacted in the global economy and to consider who is securing control of these resources and on what terms, and this book is the end product of that investigation. My hope is that it will illuminate for the reader some of the complex ethical, political, and economic implications of this new trade bio-information and, in so doing, contribute to our understanding of how new technologies will transform our relationship to the natural world in this rapidly evolving, techno-scientific age.

TRADING THE GENOME

1. INTRODUCTION

Late one Thursday afternoon in February 1996, I arrived in the sleepy town of Frederick in the heart of rural Maryland. As I stepped out of the car, my escort directed me towards a wholly unremarkable brown warehouse. It was a bright afternoon and, as we entered the facility, it took some moments for my eyes to adjust to the light, but as they did, I began to discern the familiar shapes of bulky machinery, equipment, and storage containers. Or at least they seemed familiar. But as my guide drew me onward, he explained that the units looming overhead and humming quietly under the florescent lights were not, as I had imagined, simple stock containers, but in fact, twenty-eight double fail-safed, double-decked, walk-in cryogenic storage chambers. Languishing in a suspended state in this technoscientific ark were more than 50,000 different samples of plants, animals, fungi and other organisms. Some of the materials would have been familiar to Noah—whole frozen starfish and yew leaves, for example—but others—thousands of tiny vials of extracted genetic and biochemical materials—would not.

What were these materials doing in the rural backwater of Frederick, Maryland? Where had they come from, who owned them, why had they been collected, and why were they being protected so assiduously? It was my desire to find answers to these questions that had bought me to this uncanny place. This particular voyage of discovery had begun for me some years earlier. Like many other researchers in the physical and social sciences, I had become fascinated, during the early 1990s, by the potential applications and implications of the development of new biotechnologies. As an economic and cultural geographer, I had long been interested in the creation, operation, and evolution of markets in materials or commodities that seem to stand outside any conventional notion of what could, or should, be tradable—carbon emissions, insurance risks, body parts, or child labor, for example. New biotechnologies, it seemed to me, could create not only new biological entities of an unprecedented kind but also vibrant new markets for those entities.

At much the same time, I also began to read about a series of new international biological-collection programs, "bio-prospecting projects," as they were called in the press. "Bio-prospecting" was defined by Reid et al. in the early 1990s as "the exploration of biodiversity for commercially valuable genetic and bio-chemical resources."[1] These projects were being undertaken in many tropical countries, and there was much discussion in newspapers and journals of how they would be organized, of what would be collected and where, and of whether the collection process would yield appropriate returns for supplying countries and communities. This latter question had become particularly pertinent following the ratification of the Convention on Biological Diversity of 1992, when signatory nations became obliged to ensure that all suppliers of genetic and biochemical resources receive "a just and equitable" share of the benefits arising from their utilization. Assessing whether this had happened, or was likely to happen, would, it seemed to me, necessitate establishing where the resources collected under these programs *had gone to* and how they had subsequently been used. What, in other words, had been the fate of these collections?

In answering that question, I had to consider, first, why the collections had been created. Why had there been this sudden increase in bioprospecting programs? I began by considering how the rise of this industry might be linked to other changes that had occurred in recent decades, particularly changes in the broader economy. This new industry seems to have had its genesis in a number of important technological and economic changes that occurred in the postwar period. As the world has moved from the industrial to the informational age, the economies of many of the core, industrialized countries, including the United States, have undergone a fundamental transformation. Heavy manufacturing, which once dominated economic production in advanced economies, has declined in importance, outstripped by growth in financial transactions, foreign investments, telecommunications, and other information-based services. It has been estimated that the industrialized economies' share of the world's manufacturing output dropped from 95 percent to 80 percent in the period from 1953 to 1995, while the United States' relative share of world manufacturing production decreased from 40 percent in 1963 to a mere 27 percent in 1994.[2]

It is clearly evident, however, that not all forms of manufacturing are in decline. At the same time that we were witnessing the waning of heavy industrial manufacturing, we were also seeing the birth of many new, high-tech manufacturing industries. Such industries are generally understood to include those devoted to the production of computer software, aerospace and defense components, semiconductors, satellite technologies, and the like. However, they also include an industry devoted to a somewhat different type of high-tech manufacturing—the manufacture of life. The biotechnology or life-sciences industry also produces important commodities, but they are not the familiar products that are associated with manufacturing of old such as ships, steel, or textiles. They are products that are quite alien to us—transgenic organisms, cloned animals, and artificially generated biochemical compounds that have no parallel in nature—entities that are, in effect, a fusion of the organic and the technical.

We need only glance at any newspaper to be reminded of the transformative capabilities of this new industry. Every day, it seems, we are inundated with reports of new technoscientific developments: the creation of genetically engineered foods such as the Flavr Savr Tomato, which decays at half the rate of a normal tomato; of herbicide resistant crops; and of transgenic organisms such as DuPont's patented Oncomouse, an experimental mouse genetically engineered to reliably acquire cancer within a month. In California, genes for luminescence found in deep-sea marine life have been combined with regular lawn seed to create a "glow in the dark" lawn, which, the manufacturers suggest, may have important applications as a form of security lighting.[3] It is difficult to underestimate how rapidly the biotechnology and life-sciences industries have grown in recent years and what an important new source of economic productivity they are. In 1999 alone, the U.S. Food and Drug Administration approved twenty-two new biotech drugs and vaccines, while more than 350 additional drugs and vaccines were in late-stage clinical trials.[4] It has been estimated that the top five U.S. global biopharmaceutical companies now earn in excess of nine billion dollars in revenue annually.[5] The stock market's NASDAQ Biotech Index soared 102 percent during 1999, exceeding the 86 percent increase of the NASDAQ Composite Index, and rivaling the 184 percent jump recorded at the online Internet index TheStreet.com.[6] Despite the high risks and "burn-out" of many small ventures, it appears that the biotechnology industry is set for further rapid expansion. The Biotechnology Industry Association recently confirmed, for example, that the global biotech industry is still growing by 15 to 20 percent annually.[7]

The "reinvention of life" for commercial or industrial purposes seems set to become one of the most important and lucrative businesses of the twenty-first century, and the biotechnology/life-sciences industry will necessarily play a central role in that project. But how much do we actually know about how this industry functions? Take, for example, the question of raw materials. All industries require raw materials—the shipbuilding and textile industries have, for example, long extracted steel and fiber from locations across the world, creating new resource economies that have progressively shaped the economic, political, and social fortunes of source countries. There has been a growing, if peripheral, awareness that the life-sciences industry uses samples of plant, animal, and even human genetic materials as industrial commodities and raw material, and yet there has been surprisingly little discussion of where those materials have come from, who has rights in them, how they might lawfully be used, and who, if anyone should benefit from their exploitation.

The biotechnology industry has already demonstrated that it has a voracious appetite for novel biological materials that might form the basis of new proprietary products. It was precisely because demand for such materials so outstripped available supplies that large corporations, public and private research institutes, and small entrepreneurial companies began to implement new biological collection, or bioprospecting, programs in the late 1980s and 1990s. The materials that I found sequestered in the Frederick repository are the fruits of one such collecting program. Although it is one of the largest in the United States, this facility is, however, just one of many similar collections generated in recent times. Unbeknownst to most, in the quiet confines of boardrooms, laboratories, and warehouses, away from the public gaze, executives, scientists, and technicians have been involved in a worldwide collection project unrivalled in scale and scope since colonial times. A surprisingly diverse range of organizations, institutions, and entrepreneurs from large, privately funded pharmaceutical companies, publicly funded museums of natural history, small biotech start-ups, and even individual brokers have all been actively involved in systematically amassing, archiving, and storing hundreds of thousands of samples of genetic and biochemical materials extracted from species of plants, animals, and fungal and microbial organisms drawn from hundreds of localities around the globe.

Given the scale of the enterprise, it is surprising that so few people are aware of the existence of these collections and that so few questions have been raised about how they are being used. Most of the collections that have been created in recent times, including those created by publicly funded institutions such as museums of natural history, have been drawn into the service of industry. This is not necessarily problematic, as long as some proportion of the profits that are generated from this commercial use are redistributed to suppliers in accordance with the edicts of the biodiversity convention. The evidence produced from this study suggests, however, that this compensatory process is in danger of being fatally undermined by changes that are occurring in the way these biological materials are utilized within the life-sciences industry.

A clue to the nature of this change can be found in the wording of the biodiversity convention. In setting terms and conditions, the convention makes reference to the use of both "genetic materials" and "genetic resources." The decision to employ two separate terms suggests that there was, at the time of drafting, and that there remains a sense that genetic resources may prove to be separable from the biological materials in which they have conventionally been embedded. This is certainly so. Genetic resources may be considered to encompass genetic and biochemical material but also genetic and biochemical information derived from that material. Biotechnology has played a crucial role here, enabling the latter to be utilized independently of the former. Genetic resources may now be rendered in a variety of progressively less corporeal and more informational forms: as cryogenically stored tissue samples, as cell-lines, extracted DNA, or even as gene sequences stored in databases. When in these new artifactual forms, genetic resources become infinitely more mobile and hence more transmissible.

This has proven to be of great significance for life scientists. The molecularization of biological research has transformed approaches to the study of disease and pharmaceutical development. While researchers are still interested in examining specimens morphologically, these examinations are now, almost without exception, undertaken in concert with analyses of their genetic or biochemical composition. Although molecular level investigations may be conducted using minute amounts of biological material, they may still yield valuable quantities of genetic or biochemical information. In many cases, it is this genetic and biochemical information that is the commodity that is actually sought, collected, and valued. Scientists may thus be less interested in obtaining what the historian of science Donna Haraway calls "thick messy organisms" than they are in acquiring genetic or biochemical information that can be derived from them.[8] If this information can be made available to them in a more accessible or manipulable form, this would be, for scientists, even more convenient.

The desire and ability to translate existing resources into more transmissible, informational forms is certainly not confined to the biotechnological realm. On the contrary, it can be seen in a number of other domains: in the financial sector, in publishing, and in the music and film industries. This revolution is, in fact, in keeping with predictions that have been made about the increasingly important role that information and informational resources will play in the global economy at the beginning of this new millennium. Commentators such as Daniel Bell, Alain Touraine, and, more recently, Manuel Castells have for the past two decades argued that the saturation of the market for manufactured goods will lead to the formation of a new mode of production. This mode of production, they have argued, will rely not on industrial development, but rather on the production, processing, distribution, and consumption of *information-based* goods and services. Information itself will become an important commodity, access to and control of which will increasingly provide the basis of competitive advantage in the global economy.

"Control" is the key word here. Information-based products (such as computer software, recorded music, and electronic data) are very easily disseminated and circulated through new networking technologies, but they may also be very easily replicated. This makes the task of preventing the unauthorized reproduction of these works, and the consequent loss of income, particularly difficult to prevent. It is precisely because these informational products can be so easily circulated across and beyond existing national and international jurisdictions that they have become subject to new forms of global regulation. Global regulations have had a dual and, it could be argued, conflicting role to play in this new informational economy. On the one hand, they have been introduced in order to facilitate the exchange of information-based goods and services through the development of new markets and avenues of exchange. At the same time, they are also increasingly relied upon to restrict the unlicensed use and circulation of these same resources. Research has revealed not only the emergence of new global markets and networks of exchange for computerized and digitized information, but also the difficulties of effectively regulating these new "spaces of transmission."[9]

Translating genetic resources into new, less corporeal and more informational forms might enable them to be circulated much more rapidly around avenues of exchange, but it also makes it much more difficult to keep track of where they go and who uses them. It is possible to predict that the difficulties that other industries have faced in attempting to track or control the unlicensed dissemination, uses, and reproduction of their information-based resources might now also arise in relation to the unlicensed dissemination, use, and reproduction of these bio-informational resources. Few detailed investigations have yet been conducted into the fate of collections of genetic and biochemical materials acquired under contemporary bioprospecting programs or into the efficacy of contractual arrangements or regulatory policies designed to compensate for their use. Perhaps more importantly, no investigation has sought to question what impact the ability to render genetic and biochemical materials in more transmissible and/or informational forms might have on the effectiveness of existing compensatory regimes. My intention in undertaking this research was to remedy that situation.

As most of the genetic and biochemical materials that have been collected under contemporary bioprospecting programs have been acquired for use in the American pharmaceutical industry, it seemed appropriate to concentrate most of my attention on that industry's role in the development and operation of these programs throughout the decade of 1985 to 1995. I knew that the biotechnological revolution would radically alter the way in which genetic and biochemical materials and information were embodied and presented, and I was centrally interested in describing these changes and analyzing their impact on the dynamics of biological-resource exploitation. I particularly wanted to explore the effect that these developments could have on creating new resource economies in bio-information.[10]

In undertaking this research, my aim was not just to describe the creation of this new resource economy but also to investigate the power relations inherent in its establishment and continued operation. This entailed examining how such materials are collected and used, by whom, and under what terms and conditions. I sought to establish, first, how valuable bio-informational resources were likely to become in this technoscientific age and, second, where the locus of control over these new resources is becoming centered politically, culturally, and geographically. This demanded that I also investigate how, and by whom, new global regulatory policies relating to the commodification of genetic resources have been devised. In undertaking this analysis, I gave special consideration to the role that cultural processes have played in the construction of global regulations, as these have often been misunderstood or neglected. Although often characterized as embodying universally shared beliefs or values, global regulations prove, in many instances, to have been produced out of particular, culturally specific systems of knowledge. In a work that asks questions about power and the "sociology" of knowledge, I want to illustrate how particular knowledge systems become hegemonic over space and time, creating in the process new global geographies of power and regulatory control.

At the same time, I also want to analyze the way in which power relations are played out on a much more "domestic" scale. Studying the actions and behavior of the small group of elite actors who have acquired responsibility for organizing the operation of this trade enabled me to ascertain how the terms and conditions of exchange that govern this nascent industry were established and examine whose interests they favor. By drawing out the links between these two scales of regulation, it is possible to illustrate how global regulations have been adopted and employed at regional and local levels in order to legitimate and facilitate the development of a new resource economy in bio-information.

One of the great pleasures in undertaking this project has been the opportunity that it has given me to work more closely with historians of science and with historical geographers. They sensitized me to the importance of historicizing these biotechnological advances within the long *durée* of technological development of the eighteenth, nineteenth, and twentieth centuries. In fact, one of the most complex challenges of this project proved to be teasing out what is and is not "new" about this phenomenon. The collection of biological materials has, after all, a long history, as does the applied use of genetic and biochemical material or information—in drug development or selective breeding, for example. Nevertheless, it still seems that the recent biotechnological revolution has dramatically altered our relationship to and use of collected natural materials.

In order to ascertain how this relationship has altered, I set out in chapter 2 to examine how approaches to the collection and use of natural materials have changed over time. I was particularly interested in exploring how the values that attach to particular natural objects or materials are altered through processes of collection. What may be a relatively commonplace object in one setting may become uniquely valorized as a consequence of its removal and relocation to a different geographic, cultural, or epistemological milieu. In other words, the social construction of the value of particular materials is inextricably linked to their spatial disposition. In order to draw out these interconnections, I develop a three-phased typology of what I refer to as "the social and spatial dynamics of collecting" through an illustrative, rather than comprehensive, history of collecting within natural history. My principal argument here is that the power and profit that can accrue from being in possession of a collection of natural materials derives, in part, from three factors: first, from an ability to acquire decontextualized and therefore exoticized material; second, from an ability to concentrate and control such materials within particular localities and systems of knowledge; and third, from an ability to then recirculate or redeploy these collected materials to strategic advantage over both space and time.

I then consider in chapter 3 how new, sophisticated technologies have combined with important and related economic and regulatory changes to enable a select group of collectors to *speed up* these social and spatial dynamics of collecting—to make it easier for them to collect, concentrate and control, and recirculate and regulate valuable bio-informational resources to their personal advantage. These changes are described in turn, beginning with the technological. It was suggested as long ago as the mid-1980s that biotechnological advances change the way in which biological materials are valued and that in the life sciences, at least, these materials would come to be valued more as informational resources than as material ones.[11] Several theorists made seductive, if attenuated, references to "the information embodied in living organisms" and even went so far as to suggest that biotechnology ought to be considered as a new informational technology on the basis of its ability to decode and reprogram this genetic and biochemical "information."[12] The idea that biological materials contain important and commercially valuable genetic information has quickly entered the public domain, finding expression in the development of tropes such as "genes-as-information" and "genetic software" that are now employed routinely in many newspapers, popular magazines, and science journals.

Conscious that these informational metaphors are often employed quite loosely, I begin by critically assessing something of the history of their use in the biological realm, examining how, if at all, such terms might still be usefully employed. In making an admittedly limited case for their retention, I draw the readers' attention to the creation of new bio-informational artifacts—sequences of DNA stored in databases, for example, and the development of new and lucrative markets for these novel forms of bio-information. In explaining the emergence of this new resource economy in bio-information, I link these developments in the technoscientific realm to broader changes that have occurred in recent decades—most notably, the rise of what has been termed elsewhere "the informational economy."

The final section of this chapter is devoted to an analysis of the new regulations that govern the use of bio-informational resources, including both intellectual-property-rights regulations and benefit-sharing agreements. Here, I question what it is that these regulations seek to protect, whose interests the regulations serve, and how well they serve them. I conclude the chapter by hypothesizing that these various technological, economic, and regulatory changes are together transforming trade in biological materials, altering what is collected, how it is collected, how it is then used, and how these materials are controlled or monopolized. These factors, I argue, are combining to enable a select group of collectors to exclusively capitalize on an emergent but potentially lucrative trade in bio-information.

In chapters 4, 5, and 6, I produce detailed empirical evidence to support this hypothesis. In chapter 5, I document the extraordinary resurgence of interest in biological collecting that has taken place over the last decade, contextualizing this within a longer history of collecting within the pharmaceutical and natural-products industries. I show here that although it is possible to situate contemporary bioprospecting within an apparently seamless and unchanging continuum of biological-collection practice, it would be foolish to do so uncritically. Major changes have taken place in recent years in scientific, technological, and regulatory domains that have acted to fundamentally alter the ways in which biological materials are constructed and valued as industrial raw materials within the bioprospecting and pharmaceutical industries. This has consequently affected both *what* is collected under contemporary collecting programs and *how* these materials are collected and stored. As I reveal in this chapter, these trends have allowed certain privileged collectors to acquire, collect, and concentrate increasingly valuable bio-informational resources with much greater efficiency.

In chapter 6 I explore the fate of these collections—investigating in detail how they have been used, traded, and exchanged within the U.S. pharmaceutical industry. The materials that were collected in far-flung places and later transported to private repositories in the United States have since undergone many physical transformations, existing now in a variety of more or less corporeal forms. Each of these new artifacts—cell lines, cryogenically stored tissues samples, extracted DNA, and sequenced DNA stored on databases—have been subject to many subsequent processes of trade and exchange. Rendering collected materials in these new, more purely informational forms has enabled them to be transacted in novel ways, and I provide some first evidence of these in this chapter. I also reveal how technological advances have acted to revalue existing banks of biological materials, creating an incentive for them to be illicitly "remined" for commercial or industrial purposes. The evidence presented here suggests that these new forms of commodity exchange will also speed up the social and spatial dynamics of collecting by improving collectors' ability to successively reutilize and recirculate bio-informational resources to their further economic advantage.

As I have noted, new protocols introduced under the Convention on Biological Diversity in 1992 oblige signatory states to ensure that suppliers of genetic and biochemical resources receive "a just and equitable" share of the benefits that arise from their utilization. If this commitment is to be met, suppliers should share in the profits that accrue from the commercial exploitation of their genetic resources—*no matter how constituted.* However, for this to be possible, they must necessarily be able to trace all the uses that are made of the bio-information extracted from their collected materials, for it is this, I would argue, that is *actually* the commodity that is sought and transacted for commercial gain. As I show in chapter 3, the task of tracing the successive uses that are made of information is particularly complex. In chapter 7, I expose some of the many difficulties that are inherent in regulating transactions involving forms of bio-information, and demonstrate why a failure to successfully monitor these subsequent uses will have potentially disastrous consequences for the suppliers of genetic and biochemical resources, many of whom are groups in economically vulnerable developing countries. If this outcome is to be avoided, some new and possibly quite radical approaches to the governance of this new resource economy in bio-information will have to be implemented. Some initial thoughts on what form these approaches might take are outlined in the concluding chapter.

Before any of these issues can be dealt with, it is essential to begin contextualizing contemporary bioprospecting activities within a longer history of collecting in natural history. I do so for two reasons: The first is to remind the reader that although bioprospecting is often characterized as an activity devoted to the *exploration* of biodiversity, I would argue that it is, fundamentally, about the practice of *collecting*. The second is to highlight how collecting practices relating to the acquisition of natural materials have changed over time. In order to establish what distinguishes current collecting practices from earlier ones, it is helpful to begin by first determining what they have in common—what principles unify all types of collecting practices, both historical and contemporary.

2. THE COLLECTION OF NATURE AND THE NATURE OF COLLECTING

The collection of rare natural materials as a project for the "acquisition of the exotic" has a long and well-established history. Late-medieval princes, from Frederick II to the Duke of Burgundy, included rare and exotic natural objects among their most coveted (and private) possessions.[1] In 1514, King Manuel I of Portugal made a gift to Pope Leo X of a white elephant, "Hanno," the first to be seen in Rome since the fall of the Roman Empire.[2] The arrival of the elephant was an event witnessed by countless thousands of spectators. Every rooftop was packed with astounded onlookers, all avidly seeking a view of the monstrous curiosity. As the procession neared Rome, their numbers swelled so much that several rooftops collapsed under their collective weight. Wealthy noblemen of the region rode long distances on horseback for a glimpse of the elephant and tried to coerce the overseer to interrupt his journey and bring the remarkable beast to their castles and villas.

This fascination with rare and exotic species of animals, plants and insects has continued unabated over the centuries. Throughout the Enlightenment, the collection and examination of plants and animals became formalized as aristocrats, gentlemen scholars, and, later, botanists and zoologists began a systematic study of the natural world. As familiarity with local flora and fauna exhausted the potential for further revelations, voyages of discovery and exploration were instituted. By the eighteenth century, realms of the natural world that had previously remained insulated from the inquisitive gaze and acquisitive reach of "modern science" and commerce were being opened to investigation. Large numbers of overseas collectors (both professionals and dilettantes) were invited to gather species of crops, seeds, medicinal plants, and other valuable specimens and to transport them "home" in the service of empire.

The enterprise of scouring nature for exotic or commercially valuable specimens seems to belong to that now distant colonial era. However, throughout the 1980s and 1990s, a number of stories began to appear in newspapers and magazines that drew attention to a new nature hunt. Not since those days of the great colonial voyages of discovery has a project of exploration been reported with such alacrity. In fact, much of the reportage about this new nature hunt draws explicitly on expressions and imagery traditionally associated with earlier eras of discovery and exploration, as the titles of some of these articles—"Prospecting for Nature's Chemical Riches,"[3] "The Gene Hunters,"[4] and "Antifungal in the Jungle"[5]—suggest. The articles develop the association further, drawing upon a familiar cast of characters and locations. The stalwart scientist or bold entrepreneur is there, enduring inhospitable conditions in search of new finds, as is his native companion who directs and assists him—albeit with some surprisingly sophisticated technology: "the doctors and ethnobotanists at Shaman Pharmaceuticals tend to wear jeans and hiking boots and hump through the rain forests of Ecuador or Thailand's Golden Triangle. . . . They look for medicinal plants by following a native healer through the rain forests, videotaping the root he recommends for a cough or cutting a leaf or stem he says soothes a fungus."[6]

Journalists are not alone in drawing upon the discursive tropes and imagery of the colonial era. Materials collected under these contemporary bioprospecting programs are routinely transferred from the "developing" to the "developed" world and there commodified and patented, which has prompted commentators such as Vandana Shiva and Pat Roy Mooney to argue that these collection programs not only mirror earlier colonial projects of appropriation but in fact constitute a new and even more pervasive form of "biocolonialism" or "bio-imperialism."[7] Those responsible for undertaking bioprospecting programs are, however, at pains to refute such accusations. Almost the entire corpus of literature produced by those actively involved in contemporary bioprospecting has been devoted to emphasizing the important distinctions between historical and contemporary collecting practices. Principal amongst these is the argument that while contemporary collecting programs may still be acquisitive, they are no longer exploitative. Proponents of contemporary bioprospecting programs argue that they afford new economic opportunities for both recipients and suppliers and can consequently be understood as vehicles for preserving or salvaging biological and cultural diversity.

Although "bio-imperialist" and "bio-salvage" arguments are enticing, even persuasive on a first reading, both are highly rhetorical and, like other generalized claims, made opaque by their lack of attention to detail and geographic and historical specificity. My intention in investigating the fate of contemporary collections of biological materials is not to provide evidence for one or the other side of this now highly polarized debate, but rather to begin to question and to reformulate the very parameters within which this debate is constructed. There are undoubtedly some parallels between the ways in which biological materials have been and are now collected, controlled, and exploited; however, there are also some very important differences and discontinuities. Part of the risk associated with conflating understandings of historical and contemporary collection programs lies in reducing the complexity of the current situation by creating a comparison that is insensitive to the changes that have occurred in collecting practices over time.

In order to understand the nature and significance of the changes that are currently occurring in the collection and exploitation of natural materials, it seems essential to begin by situating these practices within a longer history of collecting within natural history. A striking feature of much of the literature about contemporary bioprospecting is the emphasis that it places on the notion of "searching." Many recent contributions have made references to "examining the plant kingdom," "exploring biological wealth," "making systematic searches," and the like.[8] Although such references invite the reader to characterize, perhaps even valorize bioprospecting as an activity dedicated to exploration and discovery, bioprospecting remains, like many earlier projects of "exploration and discovery," fundamentally about the practice of *collecting*.

REVEALING THE SOCIAL AND SPATIAL DYNAMICS OF COLLECTING

"Collecting," which has been variously described as "the appropriation of exotic objects"[9] or, more broadly, as "the gathering together of chosen objects for purposes regarded as special,"[10] has long been central to the practice of natural history. Within this context, collecting has usually been understood as either a benign, aesthetic pastime or as a dispassionate scientific venture, but rarely as an inherently political activity.[11] Yet collecting can be understood politically, as it is a process that enables individuals or groups to alienate particular bodies of material for their exclusive use. Despite this, little attention has been devoted to exploring the dynamics of collecting or to examining the ethical, political, economic, or cultural implications of contemporary collecting practices. In order to fully appreciate these implications, it is important to begin by developing a more comprehensive understanding of what it is to collect, why collecting programs are instituted, how collections acquire their value, and the power relations that are inherent in many collecting practices.

To achieve this end, I believe that it is necessary to examine both the so-cial *and the spatial* dynamics of collecting—processes that are, I would ar-gue, mutually constitutive. The social dynamics of collecting—the ability of particular groups to access, acquire, concentrate, and monopolize materi-als—are more immediately apparent to the reader and involve questions of power, privilege, opportunity, and desire. The spatial dynamics of collecting are perhaps less obvious. Much has been made in recent times of the role that particular spaces, such as museums or botanical gardens, play in col-lecting practice.[12] Here however, I hope to broaden understandings of how space is implicated in collecting practices by suggesting that it is not just spaces per se but *spatial relations* that underpin the dynamics of the col-lecting process.

Spatial relations—the relationship of objects to one another, to locality and in circulation—play a central role in the mechanics of collecting and thus in the inherently political process of annexing and monopolizing spe-cific collections of material, although this has not always been explicitly rec-ognized either generally or in existing literatures on collecting practices. Collecting is often perceived as being about an initial appropriation and transference of individual objects. However, this is perhaps best thought of as simply the first act in what proves to be a complex *process* of collection, which not only entails the acquisition but also the concentration, disciplin-ing, circulation, and regulation of assemblages of material. To gain a more nuanced understanding of the politics of collecting, it is necessary to look beyond initial acts of appropriation and transference and begin to consider how the value of collected materials can escalate with their movement over time and space, and to question who ultimately enjoys the benefits of these progressive revaluations.

I argue here that spatial relations are particularly evident in three distinct phases of the collecting process: first, in the very act of acquisition, through processes of mobilization and decontextualization; second, in the way col-lected materials are concentrated and controlled within particular localities; and third, in the later recirculation and regulation of the flow of collected materials around circuits of exchange. In the following sections of this chap-ter, I analyze each of these three phases of the collection process in turn, drawing on illustrative examples from what is, inevitably, a fragmentary rather than comprehensive history of collecting practice.

COLLECTING AS SIMPLE ACQUISITION:
DECONTEXTUALIZATION AND EXOTICIZATION

Spatial relations are implicated in collecting practices in various distinct but interconnected ways. They are implicated in the very way in which "collecting is" defined. Collecting is "a bringing together," a process that necessarily involves transfer across space.[13] At its simplest, the value of a collection derives from the fact that this process involves removing material from one place and relocating it to another. This process of decontextualization *alone* may serve to exoticize the objects in question and confer on them a particular value. "Hanno," for example, was chosen as a gift for Leo X with the intention that he might form part of the pope's famed menagerie of collected animals. The menagerie included spotted leopards, panthers, apes, lions, and even "a papal parrot in a cage."[14] Each had been selected for its rarity or exoticism and either purchased directly by Leo from enterprising collectors or proffered to him as gifts that might suitably reflect the importance of the pope and the esteem in which he was held by foreign heads of state. Hanno (together with his papal owner) were later the subject of various reverential poems, odes, and dedications.

The animals in Leo's menagerie were considered extraordinary only because they were alien to the local population. Even animals such as elephants, lions, and parrots are relatively mundane in their usual habitats. However, when placed in a different context, Hanno and the other collected animals became exoticized and emblematic of wealth and power: wealth, as such objects were difficult to obtain and thus expensive, and power, as only those in positions of privilege had the means or influence to acquire them. As the historians Lorraine Daston and Katharine Park suggest, many wondrous or freakish objects such as elephants were also thought to be imbued with magical or protective qualities. In such cases, collectors found themselves in command of an even greater power: the ability to either "restrict access [to this supernatural object/power] to certain chosen individuals, to unleash it on the unwary or monopolize it for himself."[15] In these machinations, are evidenced the genesis of what might be termed "a politics of collection and circulation."

The museologist Susan Pearce provides a more contemporary example of the way in which recontextualization can act to transform the values that attach to particular objects. In *On Collecting*, she describes the case of a young English woman who collects ordinary building bricks, stacking them around her house.[16] The bricks that she collects come from a variety of building and demolition sites. When in these conventional settings they remained comparatively valueless. However, once recontextualized, that is to say, resituated within and related to a collection of other similar objects (in this case other types of building bricks), they acquire a special and further value that exceeds that which would normally attach to an object of this type. A process of de- and recontextualization has occurred here, even if the recontexualization process is here more epistemic than geographic. The bricks, unlike Hanno, are prized not because they have been exoticized by their relocation from one site to another (many were only conveyed a short distance), but rather because their inclusion in a formal collection of related objects enabled these commonplace items to be recast and valorized as exemplars of a lost industrial tradition.

Anthropologist James Clifford's work provides a further illustration of the way in which processes of de- and recontextualization can serve to revalue objects. In *The Predicament of Culture* he recounts the story of the surrealists' interest in a collection of magnificent Eskimo masks that were languishing in the warehouse of the Museum of the American Indian in the Bronx in the mid 1940s. The museum's director, George Heye, called them "jokes" and sold half of them to the surrealists for thirty-eight to forty-five dollars each. The surrealists later reassembled these artifacts and other pieces from private collections to form an exhibition that went on display at the Betty Parson's gallery on Fifty-seventh Street. As the anthropologist Edmund Carpenter noted, by simply *"moving the museum pieces across town, the Surrealists declassified them as scientific specimens and reclassified them as Art."*[17] The masks were revalued commercially as a direct consequence of this process—they were later sold for considerable sums of money when circulated as objets d'art within an emergent and flourishing market in "primitive art," their newfound value "guaranteed by their 'vanishing' cultural status."[18]

Both Pearce's and Clifford's accounts serve to remind us that although processes of decontextualization and exoticization *are* usually dependent on movement across space, the movement need not be over large distances. The Eskimo masks had initially been revalued by virtue of their relocation from one country and cultural context to another, but more latterly by their relocation from one institutional setting to another, over a distance of not more than a few miles. The examples serve to illustrate that it is not the distance itself, but rather the degree or type of decontextualization that occurs (either geographic or epistemic) and the relative novelty of that process, that serves to transform economic and cultural relationships to collected objects.

These upper New York City art galleries might be paralleled with the "cabinets of curiosities" that became so popular in European society in the early seventeenth century. The objects contained in cabinets of curiosities were, like the masks in Betty Parson's gallery, valued for their rarity or peculiarity as well as for their ability to stand "metonymically, for a whole region or population."[19] At a time in the late seventeenth century when the discovery of foreign lands, European population growth, and new technologies began to fuel the rise of capitalism and heighten interest in colonial expansion, "collecting the world" became a new compulsion.[20] Collectors desired "not only to know the world but to know it as an enchanted place."[21] One way of achieving this was to bring together objects from all over the world so that the whole wonder of creation might be taken in at a single glance. Exotic natural or cultural objects—especially marvelous or fantastic curiosities, rarities, and "freaks" of nature—held a particular fascination. A visitor to the London home of the amateur naturalist and adventurer Walter Cope at the end of the sixteenth century described some of the array of objects to be found in his cabinet of curiosity:

> An African charm made of teeth, a felt cloak from Arabia, . . . the twisted horn of a bull seal, an embalmed child or *mumia*, an inscribed paper made of bark, a unicorn's tail, a flying rhinoceros [sic] and flies of a kind that "glow at night in Virginia instead of lights, since there is often no day there for over a month."[22]

As this account suggests, these collections were often encyclopedic—the product of an indiscriminate, covetous desire for all manner of curiosities. Although these assemblages were broadly categorized as either *naturalia*—such as minerals, plants, fossils, and stuffed animals—or *artificialia*—weapons, paintings, artifacts, and the like—and displayed with the intention of producing a pleasurable symmetry—the giant dinosaur bone placed artfully next to the minute bat bone—they were not characterized by a systematized approach to either accession or display.[23] In fact, a systematic approach to the classification of the materials would have resulted in one object being surrounded by others of a similar or only slightly different variety—thereby defeating the collectors' purpose: to juxtapose the most diverse objects in the collection in order to illuminate the contrasts between them and, in so doing, to induce a sense of awe and wonder.[24]

Rather than being collected and valued for their relationships to each other, the objects within these cabinets enjoyed a certain sovereignty, with each valued independently as a self-contained microcosm of a wider, undreamt of universe. It was not until the Enlightenment that assemblages of natural material came to be elaborated into sets of formally related categories, to acquire a value that was not determined self-referentially. Nor was it until the Enlightenment that the "appropriation of the exotic," the culture of collecting, began to acquire the status of a science.

COLLECTION AS CONCENTRATION AND CONTROL
Cycles of Accumulation

The Enlightenment project of studying and understanding nature was underwritten by a conviction that "nature in being known, may be mastered, managed and used in the service of human life."[25] Within this epistemological framework, the utility of a collection was of the essence. The potential utility of a collection was, in turn, increasingly determined by how systematically it had been ordered. Within this new context, "the happy randomness of many of the earlier virtuosi, with their bowerbird collections of curious objects which had no apparent relation, were of little use."[26] Advantage no longer derived simply from accumulating exotic, singular objects, but rather from determining and explicating the *connections or correlations* between related materials.

The movement for greater sophistication and a more rigorous approach to the study of nature had been gaining momentum since the 1490s as part a wider reform of the field of *materia medica* (the study of nature for medicinal purposes).[27] As Findlen has noted, during this period Renaissance scholars began to critique relevant classical works such as Pliny's *Natural History* and Aristotle's *De Motu Animalium* and, in so doing, set about marshaling fresh evidence to support their respective arguments. Whereas scholars would once have relied solely on a reexamination of classical texts to either uphold or discredit existing theories, they now began to consider a new form of inquiry: the examination of nature itself. As the historian Paula Findlen explains: "as collectors became more comfortable with the idea of studying nature as material object rather than textual subject [so] they began to give equivalent weight to the activities surrounding the gathering of nature."[28]

Herbalists and apothecaries had, for centuries, amassed collections of plants, roots, flowers, marine creatures, and insects to form pharmacopoeia; however, these collections were created for applied use. They were neither comprehensive nor systematic and were, moreover, usually composed exclusively of locally derived flora and fauna. Fieldwork, which provided new opportunities for the more intensive and extensive *study* of nature, was formally introduced as an element of student training in 1540 by Luca Ghini, founder of the academic study of biology at Bologna and Pisa. Although interest in the study of nature was escalating rapidly in this period, fieldtrips such as these were initially focused on the immediate environment. The naturalist Ulisse Aldrovandi, for example, despite possessing of one of the most encyclopedic collections of natural specimens in the Italian renaissance, confined his excursions to the immediate vicinity of Bologna and rarely ventured further than Verona in the north and Rome in the south.

It was not long, however, before collectors began to transgress the boundaries of local experience. The introduction of a new medical curriculum that emphasized the importance of studying *all* of nature provided a fresh incentive to expand the geographical horizons of inquiry. Collectors and student naturalists began to venture forth to previously unexplored regions of the world—the Orient, the East and West Indies, and the Middle East—in search of novel plants and animals (see figure 2.1). By the end of the sixteenth century, "travel had become an essential rite of passage for the aspiring naturalist."[29] As Findlen reminds us, though, the best collectors were not always the most well traveled. Many lacked the financial resources to fund transcontinental expeditions and had to rely on agents who could supply them with exotic specimens from abroad. Ironically, it was through this vicarious process of accumulation that these "stay-at-home collectors" came to acquire an integrated understanding of the collected materials and a capacity for "critical synthesis" that others did not enjoy. It was in this moment that interest in simple observation became translated into a new preoccupation with scientific study, classification, and categorization.

The expeditions and voyages of discovery that induced and facilitated the development of this new science would come to dominate collecting practice for the following two centuries. As the historian of science Bruno Latour suggests, the great European voyages of discovery of the seventeenth and eighteenth centuries were undertakings dedicated to both investigation and accumulation.[30] Each successive voyage amassed not just an undisciplined array of trophies or souvenirs but a systematically organized body of information about the coastlines, flora, fauna, language, and cultures of distant peoples. This information could be employed to recreate within particular dedicated spaces in Europe—such as museums and laboratories—a scaled-down version of the world that could be surveyed panoptically. By these "extraordinary means" it became possible for collectors to "see" distant places, events, and cultures without actually experiencing them firsthand.[31]

Fig. 1

Traje e utensilios para herborisar-se

Fig. 2

Lata de herborisação

2.1 Venturing forth: The early collector and his "utensils." Source: Caminhoa, *Memoria sobre o modo de conservar as plantas*, 1873. Courtesy of Kew Gardens Library.

By *concentrating* information and specimens in one location, it also became possible for collectors to gain an "overview" of *the relationships* between collected materials that would otherwise have been all but impossible for a single person in a single location to obtain. As Latour suggests, it was at this point that something akin to a "Copernican Revolution" took place: "those who were the weakest because they remained at the center and saw nothing started becoming the strongest, familiar with more places not only than any native, but than any traveling captain as well."[32] To use Latour's term, a "cycle of accumulation" was established as a consequence.[33] Armed with "foresight" (a familiarity with events, places, and materials that could be acquired prior to contact), future emissaries were able to refine their searches, search more effectively, and, thus, further the process and cycle of accumulation. A way had been devised to effect "domination at a distance."[34] However, as Latour argues, for this "mobilization of the worlds" to take place, the material in question (both organisms and information) needed to be made "stable" and "combinable."[35] It was *necessary* for it to be stabilized so it might withstand transference from the periphery to metropolitan institutions in Europe without further corruption, distortion, or decay, and *desirable* that it be combinable, such that a further advantages might be obtained from its aggregation.

Technologies of Exploration and Exploitation: Mobilizing Nature

"Accumulating the world" proved a complex undertaking, one that necessitated finding new means of stabilizing natural objects and phenomena so that they might be "bought home" and examined in greater detail.[36] However, this task of collecting and bringing home particular phenomena was often problematic. This was particularly so when the phenomena itself was either not transportable (e.g., a coastline), not transportable in its entirety (e.g., a mountain), or not easily transportable (e.g., a large wild animal). Even in instances where it was possible to capture nature, it rarely proved easy to convey. During the earliest voyages of discovery, plants and animals were collected as live specimens and transported in tubs or cages, but many died in transit, unable to withstand the dramatic changes to their environment. It was clear that new ways had to be found to bring these objects home without bringing them home either "for real or for good." It was, after all, often both impossible and undesirable to bring these phenomena in for real: they were too cumbersome or too numerous. Transporting them in their original form—even if it were possible—would only result in home institutions being flooded with all manner of unwieldy phenomena. Resolving the paradox of how to bring something home without bringing it home for real, or for good, involved trying to devise ways of reducing the phenomena in question to a more readily transportable, *elemental* form.

This goal was realized in the seventeenth and eighteenth centuries through the development and introduction of what Latour refers to as "inscription devices." These were new technological instruments—astrolabes, greenhouses, marine clocks, navigation charts, nibbed pens, taxodermic tools, and the like—that provided new, more sophisticated means of "reeling in" nature—of bringing natural entities, *or at least some of their key properties*—home.[37] These novel technologies facilitated this process by enabling nature to be translated into new, more artifactual forms (specimens, botanical illustrations, stuffed animals, and so on) that might effectively "stand in for" or "represent" the organism in question. They released collectors from the obligation of keeping whole specimens alive over great distances and periods of time by generating "proxies": parts of organisms or representations of them (grafted tissues, collections of seeds, herbarium sheets,[38] and botanical illustrations) that might prove sufficient for purposes of classification or even propagation.[39] They provided a further, innovative means of decontextualizing, fragmenting, extracting, fixing, and thus *mobilizing* notoriously unstable natural materials (see figure 2.2).

Most of these inscription devices were portable and designed to be employed "in the field"—to be taken *to* the phenomena or specimen requiring capture. Even large devices such as herbaria were initially designed to be "mobile," and many were carried aboard ships of discovery.[40] However, as these herbaria grew in size, they became too unwieldy to be transported easily. Consequently, they were re-embedded within those institutions that had become focal points for such cycles of accumulation—museums, laboratories, and botanical gardens. Although the latched whitewashed cupboards of institutional herbaria remained strongly reminiscent of those once found below the decks of the *Endeavour* and the *Beagle*, their traveling days were over. The polarity had been reversed—the herbaria would no longer be taken to the specimens, the specimens would now be taken to the herbaria.

Greenhouses were another new technology in the 1700s, and they also played a pivotal role in the mobilization of nature. Unlike other inscription devices, greenhouses were unique as they enabled the mobilization not only of specimens but of *entire environments*. Until their introduction, environments had formed an integral and inextricable part of the constitution of particular *places*—it was impossible to experience one in isolation from the other.[41] The invention of the greenhouse enabled the "specificities of place"

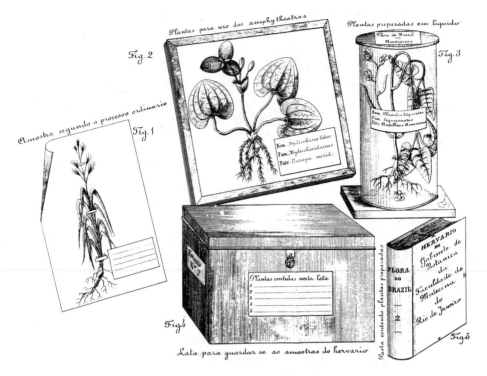

2.2A Early bio-informational proxies. Botanical illustrations, herbarium specimens, fixed specimens, florilegium.

so necessary for the propagation and sustaining of particular plants to be unproblematically reproduced in any given location (see figure 2.3).[42]

It is important to remember, however, that this capacity was limited in two important respects. First, only environmental *conditions* were reproduced in greenhouses, not *entire ecosystems*. Many sensitive, context-dependent, and symbiotic relationships that had evolved between plants and other organisms in their wider environment were either disrupted or entirely destroyed in the course of their transfer to the greenhouse. Although not as radically stabilized as their counterparts now glued to herbarium sheets, greenhouse plants were nonetheless still decontextualized—extricated from the tangle of their complex natural environments—concentrated, isolated, controlled, and able to be mass-produced on demand.

Second, while it had become theoretically possible to reproduce environments and plants in any location, in reality it was only possible to repro-

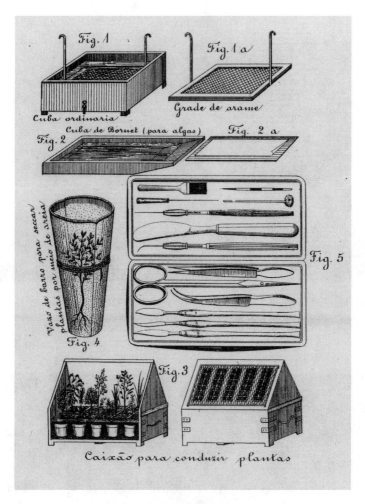

2.2B Some "inscription devices" employed in their manufacture. Source: Caminhoa, *Memoria sobre o modo de conservar as plantas*, 1873. Courtesy of Kew Gardens Library.

duce them in locations where greenhouses were to be found. The spatial distribution of greenhouses at this time is significant and should not be overlooked. As a new, sophisticated, and expensive technology, they were primarily concentrated in what Latour refers to as "the centers of calculation"—those institutions such as museums, laboratories, or botanical gardens that had become focal points for accumulations of collected specimens and data—or at their satellite outstations located at strategic points

2.3A Herbaria and greenhouses at Kew: New technologies for stabilizing and mobilizing nature. The greenhouse at the RBG, Kew, in which the seeds of the rubber plants were raised in 1876 for shipment to Ceylon. Source: Desmond and Hepper, *A Century of Kew Plantsmen*, 20, 29.

2.3B The herbarium at the Royal Botanical Gardens, Kew, opened in 1877. Source: Desmond and Hepper, *A Century of Kew Plantsmen*, 20, 29.

across the globe. These "centers of calculation" enjoyed exclusive access to these new technologies and were able to exploit them to great effect in the much wider imperial project of collecting and redeploying plant resources to economic and political advantage.

Systems for the Classification of Natural Materials

Although the task of stabilizing and mobilizing nature was demanding, it was not as complex as finding an appropriate means of stabilizing and mobilizing *information* about the collected specimens. Bodies of knowledge can also suffer corruption, distortion, or decay during processes of collection and transference. If knowledge about the relationships between collected specimens was to be accurately transferred and combined, that knowledge could no longer be subjectively determined; it needed to be justified within particular terms of reference supplied by a stable, universal framework to which all contributors subscribed, a process that demanded, as the historian of science David Miller suggests, the creation of "standard protocols for naming and depicting specimens such as were provided by systems of classification."[43]

An interesting contemporary debate surrounds the objectivity of processes of scientific collecting and classification. For Pearce, the dynamics of conventional collection and display are usually subjective: "the significance of a collection lies in the *perceived* relationships between objects," with these reflecting the desires, experiences and predilections of the collector.[44] Which objects are chosen for display or discarded, how the arrangement is prioritized, and the aesthetics of this ordering all reveal the existence of an emotive relationship between the collector and the objects themselves. For Clifford, this is a relationship that is mediated by wider societal norms. He suggests that "the inclusions in all collections reflect wider cultural values— of rational taxonomy, of gender, of aesthetics . . . thus the self that must possess but cannot have it all, learns to select, order, classify in hierarchies—to make 'good' collections."[45] The value in these aesthetic or cultural collections lies precisely in their subjectivity, in the relationship that they construct and reveal between the collector and the collected objects.

In contrast, scientific collecting has been understood as an enterprise that is inoculated from the vagaries or compulsions of fetishism by the fact that it is situated within a "rational," "ordered," and "objective" scientific framework. Sociologists of science have argued, however, that scientific paradigms, models, and methods that make some claim to objectivity are no less immune to processes of negotiation, contestation, or contrivance, suggesting that "the basic descriptors of scientific activity—experiment or discovery or observation—are equally the product of complex processes of interpretation based in particular cultural practices."[46]

It is instructive, when examining the cultural embeddedness of scientific systems of classification, to take as a case in point the development of Linnaeus's system of biological classification. Historian Londa Schiebinger's research is useful here, as it reveals the degree to which Linnaeus's system was informed not, as might be imagined, by rational and objective criteria but rather by the societal and cultural mores that dominated life in eighteenth-century Europe. Linnaeus was one of the first botanists to identify that plants reproduce sexually. He drew direct comparisons between male and female reproductive parts and processes in animals and plants, and used these as the basis of his system of classification. However, as Schiebinger suggests, "he did not stop with simple definitions of maleness and femaleness . . . not only were his plants sexed, but they actually became humans; or more specifically, they became *husbands and wives*."[47] Moreover, he divided the plant world into major groups according to the *type* of marriage each plant had contracted—for example, on the basis of whether they had "wed" "publicly" or "clandestinely." Schiebinger's analysis has revealed that these traditional, socially constructed and gendered roles actually structured the taxonomy of his system. His "Key to the Sexual System," for example, was founded explicitly on this concept of the *nuptiae plantarum* or the "Marriage of Plants."

Consequently however, the system failed to capture fundamental types of sexual and particularly asexual reproduction that could not be accommodated within this gendered "cosmology."[48] The Linnaean system also severely restricted what was considered relevant to the classificatory endeavor in other important ways. For example, it "largely limited itself to a study of the external features of the organism" and even then, features such as "size, color and smell [were] largely ignored."[49] I raise these points to illustrate that Linnaeus's sexual system of classification was not normative, objective, or rational. It was, rather, a regulatory system devised by Linnaeus as he "read nature through the lens of social relations in such a way that the new language of botany incorporated fundamental aspects of the social world as he experienced it in that time and place."[50]

Linnaeus's sexual system of classification was later discredited and abandoned—but not before its use had become widespread. For while it may not have been entirely rational it was certainly accessible, and as such played an important role in the democratization of botany. Spreading via the vector of amateur naturalists and collectors, the system was successfully universalized, despite its vagaries and cultural biases. Harvard historian Lisbet Koerner has explained how Linnaeus promoted his system to students of natural history: "Linnaeus bragged that because the sexual system depended on a few, easily observable features . . . poor students and the common people could now become proficient botanists, for an expensive library and previous instruction were no longer essential to a botanist's training."[51] Flattered by the proposition that familiarity with the system might afford them the much coveted status of professional botanists, amateur naturalists began enthusiastically to adopt the Linnaean system of classification. The system became dominant in Europe during the late eighteenth century and spread quickly to the colonies via what historian David Mackay refers to as "agents of empire"—colonial collectors and naturalists.[52] It was as enthusiastically received in North America as it had been in Europe, and for similar reasons, notably its simplicity. As flawed as nineteenth-century scientists would find it, eighteenth-century naturalists embraced the system for its ability to quickly order, at least superficially, a plethora of seemingly unrelated specimens. As the naturalist Elizabeth Keeney concludes, its popularity owed much to the fact that it required no special skill or equipment and allowed the rapid "pigeonholing" of specimens. As she suggests, "this enabled the observer to get a name fast—an attribute that would ensure its popularity among amateurs and educators long after it had ceased to be on the cutting edge scientifically."[53]

Spaces for Disciplining Biological Material

Amateurs were not, however, to be entrusted with the classificatory endeavor for long. The link between field observation and classification did not persist, as most prominent naturalists held the view that "collections would provide naturalists with a source of information that was incomparably superior to any that could be gained from studying the plant in the field."[54] There is little question that more sophisticated systems of classification could not have been effected without the development of these centralized collections of material. However, the task of housing these collections generated fresh difficulties. It demanded the creation of new dedicated spaces: "highly disciplined institutional arrangements for gathering specimens together and combining them in such a way that botanists and zoologists could see new things in the wholes there constituted."[55]

It is here that the interdependent relationship between the social and spatial dynamics of collecting (and of annexation and monopolization) becomes more apparent. Hierarchical systems of classification are based on both observation (the "objective" gaze of the naturalist) and categorization.

Explicating the relationships or correlations between particular specimens was ideally achieved through *direct comparison*—an activity that demanded a concentration within a fixed locality of a number of like (probably related) objects. Comparison was not rendered impossible without spatial juxtaposition but was certainly facilitated by it. As the French philosopher Michel Foucault noted, although the establishment of botanical and zoological gardens in the eighteenth century was thought to reflect a new curiosity in exotic plants and animals, this interest had, in fact, been in existence for some while. What had changed, he suggests, "was the space in which it was possible to see them, and from which it was possible to describe them."[56]

Botanical and zoological gardens, museums of natural history, and laboratories all provided new theatres or sites devoted to the collection, concentration, surveillance, description, and disciplining of collected materials. Foucault characterizes them as "unencumbered spaces in which things are juxtaposed: herbariums, collections, gardens; [within which] *creatures present themselves one beside another*, their surfaces visible, grouped according to their common features and thus, *already virtually analyzed*."[57] The *arrangement* of this spatial juxtaposition was, he suggests, crucial to the classificatory endeavor. While during the classical period and the Renaissance exotic animals were displayed as part of a broad spectacle, "the natural history room and the garden came to replace the circular procession of the 'show' with the arrangement of things in a 'table' "—a format that aided comparison and classification. By concentrating material within one spatial frame, it had become possible for botanists and zoologists to acquire a visual and epistemic domination of most of the Earth's plants and animals without having to move more than a few hundred yards at a time.

Genesis of a Space of Flows in Biological Material

We would, however, be succumbing to an attack of spatial fetishism if we were to imagine that the built spaces *alone* had the power to invest the materials housed therein with a particular value.[58] Of course, they did not. They did, however, come to serve as important focal points for cycles of accumulation. Individual institutions, the Royal Botanical Gardens at Kew for example, became constituted as master repositories. Scientists within these institutions were ideally placed to draw upon the information and specimens concentrated therein, obtaining a mastery of them. The multiplicative effect of this process was evidenced in these institutions' being designated as "centers of expertise." This denotative change had an interesting effect on the social and spatial dynamics of collecting. Although material could theoretically be classified "in the field," it was not. Minions operating in far-flung corners of the empire were no longer entrusted to undertake classifications of materials in situ, and they began to be drawn, vortexlike, toward Latour's "centers of calculation."

When the naturalist Joseph Hooker undertook his major taxonomic studies of the New Zealand and Indian floras in the late nineteenth century, for example, he chastised foreign collectors for their inability to understand their own plants. He lamented to others that "they frequently believe species to be new when research in the extensive collections at Kew revealed them as geographic variants of a single widespread form; they have inadequate reference works; and they inconvenience other naturalists with a proliferation of local geographic or personal names."[59] Without access to the specialized knowledge and range of specimens now available in centralized institutional collections, the field collectors' purview was deemed to be limited. They were directed, in no uncertain terms, to send plants back to Kew, "the hub of the colonial scientific enterprise."[60]

Latour's "Copernican Revolution" was beginning to have its effect. Natural materials, it seemed, could only acquire value when extracted from their chaotic surroundings and subsumed within a particular system for codifying knowledge—the Linnaean system of biological classification, for example. This demanded their export to those centers within which these classificatory systems were both constructed and maintained. As these classificatory systems were predicated on observation of characteristics, accurate classification could only, it was argued, be obtained through direct comparison with other catalogued specimens. As the historian Mary Louise Pratt has argued, "Eighteenth-century classificatory systems created the task of *locating* every species on the planet, *extracting* it from its arbitrary surroundings (the chaos) and *placing* it in its appropriate spot in the system (the order, book, collection or garden) with its new written secular name."[61] Pratt's emphasis on spatial relations in this passage is not incidental. The entire project of biological classification had come to rely on the collection and transference of materials and their concentration within dedicated repositories in Europe. With these developments came the genesis of a new spaces of flows in biological material.[62]

Built spaces such as botanical gardens and museums had, by the late nineteenth century, firmly established their reputations as "centers of calculation." Here, biological materials were amassed, redefined as specimens, concentrated within collections, and disciplined within particular knowledge systems. These repositories had become focal points for accumulations of information, material, technologies, and expertise, and consequently they were increasingly celebrated as privileged sites for the production of scientific knowledge about the natural world. A question of great salience, though, often remains unanswered at this point: Of what relevance was this? What possible power or value could be derived from concentrating and controlling such materials in this way? Imagining the project at work, it is difficult to summon up anything more than a rather benign image of a few fusty boffins in frock coats "brooding over their knowledge cabinets."[63] Undoubtedly they were masters of the knowledge that they were accumulating, and yet this knowledge only seemed to have a value within the cosmology that they were also creating. It must surely have been difficult to envision what possible wider value or power could accrue to those in possession of such an eclectic and specialized knowledge. However, as Fuller points out, "the power of knowledge lies not in its mere possession, but in the range of possible uses and users for it," adding as an important addendum, "even if these are seen primarily as the generation of more knowledge."[64] Botanical gardens, museums, laboratories, and other similar institutions were now ideally positioned to exploit both the utilitarian and the knowledge values of the specimens and information that they had been systematically acquiring.

COLLECTION AS RECIRCULATION AND REGULATION

This brings us to the third and perhaps most significant of the ways in which spatial relations are implicated in collecting practice. Analysis of the dynamics of collecting practice reveals how value is compounded at each stage of the process. A certain amount of power and value accrues when objects are first moved about, decontextualized, and exoticized. A further value accumulates when they are concentrated in specific locations where they can be ordered, controlled, and disciplined and where information about the relationships between the objects can be identified and explicated. However, it is not until the third stage that these two factors are *squared* and an even greater value obtained. By applying the knowledge, technologies, and expertise concentrated within the "centers of calculation" *to* the collected materials, it became possible to produce entirely new materials or bodies of information. In the case of botanical collectors, these recombinant materials could take the form of new varieties of plants suited to particular climates or localities, or information or data about related varieties of a plant that might substitute for ones proving difficult to obtain. The collector then enjoys the power to *recirculate* these materials or information, to *redeploy them to advantage* through strategic utilization, exchange, or trade.[65]

The collectors' capacity to command the dynamics of these processes of commodity circulation should not be underestimated. As the "owner" of the collected materials, they retain the right to determine when, where, and to whom the materials will be circulated and under what terms and conditions. By withholding the materials they may create scarcities or shortages, escalating demand in the marketplace. The collector acts as a capacitor able to store and then regulate the release of collected materials as they flow around circuits of trade and exchange. Electing when and how to release the materials is a strategic decision that collectors can exploit to considerable personal advantage. The benefits to be derived from exploiting the social and spatial dynamics of collecting in this way are nowhere more evident that in the great colonial project of recirculating collected horticultural and agricultural materials in the "service of empire."

Collecting and the Wider Imperial Project of Economic and Cultural Dominance

The principal rationale for undertaking voyages of exploration and discovery during the seventeenth and eighteenth centuries was to consolidate economic and imperial expansion. Collections of natural materials were at times undertaken for the exclusive purpose of scientific study but more frequently with a view to acquiring materials that could be employed in the service of agriculture or industry. Linnaeus's motivation for devising an efficient classificatory system is emblematic of this utilitarian ethic. Daniel Miller suggests that "Linnaeus's stress upon the utility of botany, and of natural history more generally was not incidental . . . for Linnaeus, the solid foundation of national progress lay in natural history, for natural history could help guarantee a Swedish autarky." [66]

Britain similarly turned toward the exploitation of natural history to further national and, later, imperial interests. James Cook's great voyages of the Pacific presented his botanist, Joseph Banks, with an opportunity to scour distant shores for rare specimens. Rapid industrialization within England created a demand for specific fibers, dyes, medicines, and foodstuffs that could not be met domestically. Plants, animals, and crops that were deemed to have some agricultural, industrial, or horticultural utility were sought out as industrial raw materials, with the intention of reproducing them either in England or in her colonial dependencies.

The conditions under which these specimens were acquired and the fate of these early collections of botanical material gives an insight into the transition from private to public collection and ownership of this material that was occurring at that time. Despite the fact that Cook's voyages of discovery were state-funded exercises undertaken by the British Admiralty, the collection programs undertaken on these voyages were orchestrated by Banks in a private capacity. The relationship between Banks and his collecting agents was a complex one.[67] They included horticulturists, botanists, physicians, civil servants, army officers, naval and merchant officers, navigators, and explorers, many of whom were drawn into Banks's employ on the understanding that they were in the service of science. In fact, Banks often drew materials from his collection and library for nonscientific purposes—for example, to assist commercial operations such as the East India Company. This practice was, in fact, entirely reflective of the accommodating relationship that existed between scientific and corporate interests during this period. For example, overseas trading companies often gave free passage to Linnaeus's students in return for information about collected specimens.[68] Conversely, master collectors such as Banks used their connections with such trading companies to cultivate close relationships with captains and crews, who were co-opted to collect specimens on his behalf.

As Banks had collected specimens on his first expedition with James Cook in his capacity as a private botanist, they remained under his direct control. On his return to London they were removed to Banks's residence at New Burlington Street. Here they came to constitute a part of his much larger collection of catalogued specimens. This, in turn, formed part of a consortium of significant but privately domiciled collections held in England. As David Miller suggests however, it soon became evident that "the accumulation of private power in the form of private 'data banks' and collections, which co-existed alongside growing national counterparts, [created] an obstacle to the fuller realization of the combinability of information gathered from distant parts."[69] A desire not to obstruct national advancement provided the impetus for collectors such as Banks to arrange for the disbursement of their collections to institutes such as the British Museum and the Royal Botanical Gardens at Kew.

Kew itself underwent a profound transformation following the deaths of George III and Joseph Banks in 1820. Royal patronage of the gardens began to decline in this period as George III's successors proved "as indifferent to botany as they were to books."[70] Expenditure at Kew fell from £1,900 per annum in the early 1820s to £1,300 less than a decade later.[71] The institution appeared destined for progressive dissolution and was only preserved following an impassioned campaign by members of the Royal Society. The campaign emphasized the considerable benefits that the nation might derive from acquiring, examining, and propagating plants and crops that could be serviceably employed in the national project of "agrarian improvement." However, their arguments for the retention and further development of Kew were tempered by an important caveat, that the gardens be reformed—run not by or for the interests of royal patrons or private dilettantes but rather by professional scientists committed to the furtherance of formal botanical research.

Noting the contribution that Kew and her satellite outstations could make to the imperial economy, John Lindley, author of the Lindley Report on the fate of Kew Gardens, rued the fact that although "there are many gardens in British Colonies and dependencies . . . costing many thousands a year, their utility is very much diminished by want of some system under which they can all be regulated and controlled . . . there is no unity of purpose among them . . . their powers [are] wasted."[72] A National Botanic Garden could form "a center around which all these minor establishments could be arranged" and he argued "that they should all be under the control of the chief of that Garden, . . . explaining their wants, receiving their supplies, and aiding the Mother Country in everything that is useful in the vegetable kingdom." He concluded that "medicine, commerce, agriculture, horticulture and many valuable branches of manufacture, would all derive considerable advantages from . . . such a system."[73]

In the interests of formalizing collecting practices, centralizing collections, and streamlining the operation of this network of linked botanical gardens, a decision was made in 1838 to endorse Lindley's proposal that Kew be turned into a national "scientific garden." Under this arrangement, the garden would receive some £20,000 of state funding from the treasury initially and a further £4,000 annually for its continued maintenance. On the 25 June 1840, the Royal Botanic Gardens at Kew were formally transferred from the control of the Crown to the control of state and in that moment transformed from privately funded pleasure garden to publicly funded scientific research institute. This development was to mark the beginning of the trend towards the institutionalization of biological collecting programs in the West.

The institutionalized collection of botanical material continued in earnest throughout the nineteenth century, as the European powers began to establish and then expand their worldwide network of botanical gardens.[74] These institutions would play a vital role in collecting, examining, and

ordering specimens and, more importantly, in regulating the flow of collected materials as they were transferred from and to their colonial dependencies. Increasingly cognizant of their capacity to recirculate collected materials to advantage, botanists, horticulturists, and agriculturists began to
prioritize the acquisition of indigenous plants and cultivars that could be extracted from foreign lands and regrown in their territorial possessions. Particular commodities—such as rubber and tea—were consumed in great
quantities in Europe. For example, in Britain in the second half of the nineteenth century, demand for rubber escalated so rapidly that the value of im-
ports rose from £300,000 in 1854 to £1,300,000 in 1874.[75] All of this material came from Amazonian Brazil and Peru. In the mid-1870s, botanists Joseph
Hooker and Clement Markham began to devise a scheme to transfer rubber
seedlings from South America to Kew. Using the technologies of exploration
and exploitation there concentrated, it became possible to keep these specimens alive in a transitory life-world (the greenhouse) until they could be successfully redeployed. This process involved recirculating the materials to
colonial dependencies via a series of botanical gardens located in colonial
outposts. The rubber seedlings reached plantations in Malaysia via colonial
botanical gardens in Ceylon and Singapore where they were cultivated en
masse. Brazil's share of the world rubber market consequently dropped from
95 percent at the turn of the twentieth century to 5 percent in 1980.[76]

Tea, similarly, was a commodity that, while increasingly popular in Great
Britain, was ruinously expensive, as it had to be imported from China. On
the suggestion of the director of the Botanical Garden in Ceylon, tea plants
that had been collected in China were introduced to the garden and cultivated in carefully controlled conditions. In 1873, some 23 pounds of tea derived from China were produced in Ceylon and exported to Britain. Less
than twenty years later, Ceylon was exporting 74 million pounds of this tea
to Britain—some 40 percent of all the tea consumed there at that time.[77]
Botanical gardens, then, had a dual role to play, acting not only as focal
points for collecting operations but also as holding and redistribution stations for recirculated materials.[78] The colonial network of botanical gardens
formed a series of nodes in an imperial circuit of exchange, regulating the
flow of botanical materials and information over space and time.

Achieving this task required that the "centers of calculation" had the capacity to bring into their service not only collected materials but a range of
actors, institutions, and economic and political resources within and beyond
their colonial dependencies. Colonial collectors were, as Mackay suggests,
"agents of empire" in an ideological sense, responsible for introducing the

Table 2.1   Principal plantation crops, areas of origin, and areas in which plantations were
established by 1900

Crop	Origin	Plantations established
Banana	Southeast Asia	Africa, Caribbean, Central America, South America
Cocoa	Brazil, Mexico	West Africa, Southeast Asia, Caribbean
Coffee	Ethiopia	East Africa, Caribbean, South America, Central America, East Asia, Southeast Asia
Cotton	Mexico,[1] Peru[2]	East Africa, North Africa, East Asia, South America, North America, Caribbean
Oil Palm	West Africa	Southeast Asia
Pineapple	Brazil	West Africa, Southeast Asia
Rubber	Brazil	West Africa, Southeast Asia
Sisal	Mexico	East Africa, East Asia, South America
Sugar Cane	Southeast Asia	East Africa, North Africa, Southern Africa, Caribbean, Central America, South America
Tea	China	East Asia, Southeast Asia, East Africa

1. Upland Cotton
2. Sea Island Cotton
Source: Kloppenburg, *First the Seed*, 155.

principles of naturalization, systematization, and rationality that would provide the foundation of the imperial project of ordering the economies, histories, and cultural practices of other peoples. David Miller contends that
"the first task of the collectors was to transcend the chaos of the lands they
visited and reduce the natural world of empire to order. This work could
only be done by Europeans, [it was argued,] since it was generally agreed
that science and learning in new lands were at best obscurantist and at worst
non-existent. . . . Bypassing local 'ignorance,' the collectors helped to incor-
porate new lands and colonies into a British scientific and industrial hegemony."[79] By employing such methods, European collectors were able to
identify and acquire an unprecedented amount of industrially significant
germplasm.[80] These materials were collected throughout the New World
and then transferred to colonial plantations in the Americas, Africa, and the
Caribbean for further cultivation. Table 2.1 gives an indication of the number of crops that had been transferred in this way by 1900 and of the emerging geography of this flow of germplasm.

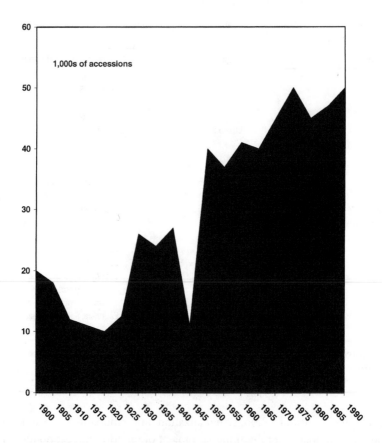

2.4 Number of germplasm accessions recorded by the USDA, by five year period, 1900–1990. Source: Kloppenburg, *First the Seed*, 160.

NEW WORLD COLLECTORS

The collection of botanical material by the European powers did not end with the collapse of empire; it continued throughout the twentieth century. In the late 1800s and early 1900s, the U.S. government became increasingly aware of the homogeneity of North America's indigenous germplasm and, consequently, of the need to institute active collection programs in more biodiverse regions. The institutionalization of germplasm collection programs began in the United States in 1898 with the creation of a special Seed and Plant Introduction section of the United States Department of Agriculture. Between 1900 and 1930, the Plant Introduction Office initiated over fifty separate USDA-sponsored germplasm-collecting expeditions—ushering in a new "golden age of plant hunters."[81] Figure 2.4 provides an indication of the number of accessions that the USDA made during this period and in the following fifty years. Accessions have risen exponentially throughout the twentieth century, falling only during the four-year period of America's involvement in the Second World War. These collected materials were primarily used in the development of new hybrids and crop varieties bred specifically for production within the United States.

The collection of germplasm and other plant materials by European and North American interests continued apace in the years from 1930 to 1980. The continued transference of materials from the colonial peripheries to the metropolitan core perhaps implies an unbroken cycle of economic exploitation. However, it could be argued that the rationale for undertaking biological collections underwent at least a partial transformation during these years. Although the symbiotic relationship between scientific study and economic advancement was still in existence, an increase in publicly funded collection programs introduced a degree of disjunction between "scientific" and "commercial" collecting. Collections undertaken by museums of natural history, botanical gardens, and other publicly funded institutions such as universities and public-research institutes were made principally to acquire specimens for classification, analysis, and display. Research on plant or animal varieties or derivatives that might prove of interest to the agricultural or pharmaceutical industries *was* conducted within some universities and publicly funded research institutes; however; much of this was "blue sky" research undertaken for the advancement of science and "the common good." While many of the findings proved to have potential industrial applications, they were not specifically designed to facilitate the development of proprietary products. It was, in most instances, considered inappropriate for material collected under this rubric to be *directly* accessed or utilized by commercial or corporate interests.

This philosophy underpinned the establishment of a series of new International Agricultural Research Centers in the 1950s and 1960s. These centers were to play a central role in a global program to improve agricultural productivity. The centers, which were located in developing countries and funded by a consortium of international donors, were each responsible for breeding a particular set of crops suited to production in a particular region. As a consequence of their involvement in this project, IARCs became central repositories for collections of plant genetic resources donated by developing countries to help further agricultural research. As publicly funded agricultural-research stations, their mission was to utilize this donated germplasm to develop new plant, crop, and seed varieties suited to local conditions, with the aim of boosting agricultural productivity in developing countries. However, reports later confirmed that many of these accessions had been improperly appropriated by multinational corporations and publicly and privately funded plant breeders, who used the material to develop proprietary products that were then patented and sold back to developing country farmers as protected varieties. This development prompted commentators such as rural sociologist Jack Kloppenburg to argue in the 1980s that IARCs and other scientific collecting institutes were continuing to act "as vehicles for the efficient extraction of plant genetic resources from the Third World . . . to the gene banks of Europe, North America and Japan." He argued that they might, thus, be considered "the modern successor to the 18th and 19th century botanical gardens that served as conduits for the transmission of plant genetic information from the colonies to the imperial powers."[82]

At this point in our fragmentary history and geography of collecting practice, the trail goes cold. Although the past decade has witnessed a resurgence of interest in the collection of biological material, little information has yet been brought into the public domain about the fate of material collected during the most recent wave of bioprospecting activity. Brief reports in the media have highlighted specific collecting projects and noted some of the concerns elicited by these activities. Recent escalations in the number of biological-collection programs have gone, otherwise, largely unremarked. Who is involved in underwriting or organizing these contemporary collecting projects? Where are these new materials being transferred to, how are they being concentrated and "disciplined," where and how are they being redeployed, and, most important of all, to whose advantage? Where are the new centers of calculation, or are they the old centers of calculation?

As we enter this new "genetic century," interest in biological materials has never been greater. Cloning, genetic engineering, stem-cell research, the sequencing of animal and human genomes, and advances in molecular biology have all focused attention on our newfound capacity to "reinvent" life. The life-sciences industry has shown a continued interest in collecting biological materials that might be used in the development of new proprietary products. The pharmaceutical industry, in particular, has been at the forefront of this research, acting as a prime consumer of biological materials and a principal instigator of many new biological-collection programs. Large public-research institutes, universities, museums, a host of smaller companies, and even individual entrepreneurs have also rushed to be involved in this new collecting project.

But *what is it* that they are now collecting? This question has slipped by, largely unaddressed. The practice of collecting biological materials for commercial or industrial use appears unchanged over time. However, the decade from 1985 to 1995 witnessed the introduction of a series of profound biotechnological, economic, and regulatory changes. These changes have linked together to transform the ways in which biological materials can be used, commercially and industrially. This in turn has altered conceptions of *what it is* in biological materials that is of most value and of how that might be accessed. What effect might these changes have on collecting practices and the social and spatial dynamics of collecting—the ability to acquire, concentrate, recirculate, and regulate the flow of collected materials to personal advantage? It is to a more detailed investigation of these matters that we now turn.

3. SPEEDUP
ACCELERATING THE SOCIAL AND
SPATIAL DYNAMICS OF COLLECTING

Having devoted the previous chapter to outlining the long history of collecting within natural history, it might seem foolhardy to begin this one by suggesting that contemporary bioprospecting—this new phase of collecting—is in any way a "new" phenomenon. As the review in chapter two clearly illustrated, the project of collecting plants, animals, and other organisms for industrial, agricultural, or medicinal purposes has a long and well-documented history.[1] Set within this context, "bioprospecting" appears to be little more than a fancy new descriptor for an age-old practice. The fact that bioprospecting activities are undertaken by the same coterie of scientific, academic, and commercial interests that have been involved in the collection of natural materials since colonial times only serves to confirm this impression.

In this chapter I want to challenge that notion. Although it may appear, at first glance, that the process of collecting biological material has remained unchanged since the colonial era, I would argue that it is, in fact, undergoing a profound transformation. While contemporary bioprospectors may seem to be simply collecting plants, animals, fungi, and other organisms, they are not. They are, by their own admission, collecting something rather different: the genetic and biochemical resources embedded within these organisms. The definition of the activity in which they are engaged confirms the nature of this change. "Bioprospecting," as it is so described, is defined as "the exploration of biodiversity for *commercially valuable genetic and bio-chemical resources*."[2]

This distinction—between the collection and use of an organism, and the collection and use of the genetic and biochemical resources therein—may appear to be purely semantic and of little significance. I would argue, however, that this apparently insignificant distinction is actually of great importance. It reflects a fundamental transformation that has occurred over the past two decades in the way that biological materials are used and valued as commodities. This transformation has been induced by a series of wider economic, technical, and societal changes: the introduction of advanced biotechnologies, the rise of an information economy, and the extension of systems of regulation—particularly systems of intellectual property rights—to new geographic and epistemological domains. Genetic and biochemical materials and, indeed, *biological derivatives of many different types*, have, as a consequence of these developments, become constituted as autonomous commodities to which private rights of ownership may obtain. This, in turn, has provided the impetus to create new markets in these commodities.

Changing the way in which biological materials can be used industrially and commercially has also had profound effects on the practice of collecting—altering what is collected, how it is collected, how it is stored, and how it is utilized. In other words, these developments have also had a marked effect on the social and spatial dynamics of collecting. I suggested in the previous chapter that the power and value that accrue from collecting particular materials derive from an ability to acquire, concentrate, control, and then strategically recirculate those materials to economic or political advantage. The *mechanics* of that process are, I argued, inherently spatial. The value that attaches to particular materials may escalate when they are transferred to specific locations or sites where they can be more readily ordered, related, controlled, and disciplined. By strategically manipulating the flow of these now valorized objects around circuits of trade and exchange—by determining how and when they are recirculated and redeployed—the collector is able to further manipulate the dynamics of circulation to personal advantage.

I argue in this chapter that the combination of economic, technological and regulatory changes that I outlined above are particularly significant as they enable collectors to *speed up* the social and spatial dynamics of biological collecting whence power and profit derive. They do so by making it easier for a select few to collect, concentrate, control, and recirculate valuable biological resources to strategic or personal advantage. I believe that a new era in the history of the collection of biological materials is being ushered in as a consequence. These developments will have serious if largely unexplored practical, political, and ethical implications. My aim in this chapter is to begin to reveal the nature of these changes and their implications in greater detail.

In order to facilitate the reader's digestion of these complex and interlinked issues, I have divided the chapter into three sections. In the first, I begin by mapping out the effect that technological changes—particularly the introduction of advanced biotechnologies—have had on the ability to "render" genetic and biochemical resources in a series of progressively less corporeal and more "informational" forms: as cryogenically stored tissue samples and bio-chemical extracts, as cell lines, as extracted or even sequenced DNA. Here, I consider the argument that the desire and ability to translate biological materials into these new forms reflects a change in the way they are valued—that they are increasingly valued (within the life-sciences industry at least) for their informational content rather than their material form. I begin my analysis by questioning how the term "information" is being used in this context, considering whether it could ever be appropriate to employ it literally rather than metaphorically. This involves a necessary reflection on the historical use of informational metaphors in biology, examining how they came to be imported into the discipline and why they went on to gain wider currency.

I turn, in the second section, to a consideration of how these developments in the biotechnological realm might be related to other changes occurring in the broader economy, most notably in the rise of an information economy. I note here the increasingly important role that information and information-based products and processes will play as sources of productivity in the postindustrial economy and outline the important role that new forms of spatial organization (the development of networks, strategic alliances, and interlinkages between organizations) are playing in creating new avenues of exchange and markets for information-based products. I look in closer detail at the effect that informational technologies such as computers have had on translating existing resources into new, more purely informational forms and reveal the impetus that this has given to the creation of new markets for these informational artifacts. I then draw some important parallels between the actions of information technologies and biotechnologies, revealing how the application of the latter may similarly give rise to new markets in bio-informational artifacts. My aim, in so doing, is to explore and refine existing theories of the effect that these technological and economic developments might have on trade in biological materials.

I then turn, in the third section of the chapter, questions of regulation. Here, I investigate how new global regulatory regimes (particularly intellectual property rights regimes) have acted to facilitate the creation of new markets in biological derivatives and information by extending intellectual property rights to what might be termed "embodied inventions." Examining in detail protocols introduced under the General Agreement on Tariff and Trades (GATT), the General Agreement on Tariff and Trades Intellectual

Property Rights (GATT TRIPs)—now a World Trade Organization protocol—and the Convention on Biological Diversity, I explore the role that
these regulations have played, first, in constructing biological derivatives as
autonomous commodities and, second, in setting the terms under which
they might be used and exchanged. Changing the way in which resources
are embodied or presented may speed their circulation around the market
but may also complicate efforts to monitor or regulate their use. In assessing
the likely efficacy of compensatory agreements, I draw attention to the complexities that are associated with regulating new "spaces of transmission,"
highlighting particularly the difficulties of monitoring the successive uses
that are made of information-based products, such as recorded music and
computer software. I conclude by reflecting on the lessons that can be derived from these experiences for those charged with regulating the use of biological derivatives, including genetic information.

Throughout, I link my analysis to the threefold typology of the social and
spatial dynamics of collecting that I set out in chapter 2, for it is along these
three axes—the acquisition, the concentration and control, and the recirculation and regulation of these materials—that the effect of changes in the
status of biological materials (ontological, economic, and legal) on the mechanics of collecting practice are most evident. I begin therefore, by looking
first at the technological developments that have occurred in the life-
sciences industry over the past two decades, considering as I go the impact
that they have had on the first axis of the social and spatial dynamics of collecting—acquisition. In addressing this issue, I am not so much concerned
with assessing how new technologies have affected what types or species of
organisms are collected. I am interested here in a rather more fundamental
question: How has the introduction of new technologies *affected what is being collected*? In short, I find myself wondering: What exactly *is* being acquired these days when people collect biological materials?

RETHEORIZING LIFE FORMS: MATERIAL AND INFORMATIONAL?

As even the partial review of natural-history collecting offered in chapter 2
reveals, there has long been interest in acquiring various types of biological
materials. These materials have been valued in different ways over time.
Plants, animals, crops, seeds, and their by-products have traditionally been
valued and sought out as sources of sustenance. Some of these natural organisms (rare animals, for example) or materials (bird's feathers, rhinoceros
horns, and so on) came to acquire a singular and exceptional value during

2.

A

D E S C R I P T I V E

C A T A L O G U E

Of a very Extenſive and Capital

C O L L E C T I O N

O F

ANATOMICAL PREPARATIONS,

ORIGINAL CASTS OF THE GRAVID UTERUS,
accurately moulded from Nature;

COLOURED ANATOMICAL DRAWINGS,

AND

N A T U R A L H I S T O R Y;

Forming the Entire and Genuine MUSEUM

OF AN

Eminent Profeſſor of A N A T O M Y,
Who has declined Teaching.

*This Collection has been the Work of many Years extenſive
Practice and cloſe Application to this Study.*

Which will be Sold by A U C T I O N,

By MR. H U T C H I N S,

At his R O O M S, in *King-Street* and *Hart-Street,*
Covent-Garden,

On MONDAY, *December* the 10th, 1787, and the Twelve
following Evenings *(Sundays* excepted*),* at Six o'Clock.

The Whole will be on View on FRIDAY the 7th and
SATURDAY the 8th Inſtant.

CATALOGUES (at One Shilling each) may be had of Mr. HUTCHINS,
Nº 41, *King-Street, Covent-Garden:* Who fixes a Valuation on
Noblemen's and Gentlemen's Effects of all Sorts, in Town or Coun-
try; and the Value, if required, immediately given for the ſame.

3.1 Catalogue of one of the earliest known auctions of anatomical specimens and preparations, Covent Garden, London, 10 December 1787. Source: Wellcome History of Medicine Library, London.

the Renaissance, when they became coveted as exotic objects possibly imbued with mystical or talismanic power. Biological materials were subject to a further revaluation during the Enlightenment, when they became prized not only for their exoticism but also for the unique position that they occupied within a wider set of related objects and a wider system of knowledge about natural history. It was at this time that particular biological organisms, or parts thereof, became privileged as exemplars (either as ideal or typical examples) of their type and thus formally characterized and intentionally archived as "specimens." These formally constituted collections of specimens came to acquire a further pecuniary value. As evidence suggests, trade in collections of biological material began to escalate in the late eighteenth century, with one of the earliest known auctions of anatomical specimens and preparations taking place at Covent Garden in London on 10 December 1787 (see figure 3.1).

At the turn of the twentieth century, the study of the natural world underwent another profound transformation as new sciences such as genetics and experimental embryology allowed living material to be opened up to more detailed investigation. As the historian of science Lynn Nyhart has suggested, there is, in fact, a greater degree of continuity between the natural-history studies that took place in the seventeenth and eighteenth centuries and these new experimental life sciences than is immediately apparent. As she reminds us, explorations into the interior worlds of living organisms have been continuing apace since 1665, when Robert Hooke first employed the microscope to reveal the inner workings of plants and fungal materials (see figure 3.2). Nevertheless, there is little doubt that the introduction of these new sciences irrevocably changed understandings of how living materials are constituted; particularly since the 1970s, when the first experiments in genetic engineering were conducted, biological material has been undergoing another process of reevaluation.

With the introduction of these new biotechnological techniques (DNA extraction, identification, sequencing, and splicing and genetic cloning) and the further refinement of other existing techniques (biochemical assaying, roboticized screening, and cell culture) it become possible to effectively and efficiently disaggregate biological materials, that is to say, to reduce them to a series of constituent parts (cells, genes, plasmids, biochemical compounds, and so on). While these genetic materials have always existed, they have not, until comparatively recently, been available for use as discreet entities, as most have remained inextricably embedded within organisms. Biochemical compounds have been more accessible, but

3.2 Early explorations into the interior of living organisms: Microscopic studies of the stinging nettle, from Robert Hooke's *Micrographia*, 1665. Source: Whipple Museum of the History of Science, Cambridge.

many have proven to be difficult to extract and extremely labile. The improvements that these new technologies offered enabled components that had been previously unknown, inaccessible, or unstable to be efficiently maintained and utilized *independently* of the organisms in which they were originally produced. These biological derivatives have consequently come to be constituted as "resources" in the classical sense, as "a stock or reserve upon which one can draw."[3]

The American geographer Cindi Katz has suggested that the impetus to turn the investigative lens inward in the natural sciences has been, in part, the product of set of broader political and economic factors. Processes of decolonialization, coupled with increasing environmental awareness, have, she suggests, served to contract the "frontiers" of development by imposing both environmental and spatial limits on growth. These constraints have forced a reassessment of the capital value of nature. Nature, she argues, "is consequently now undergoing an 'involution' much as space did in the first few years of the twentieth century when planetary expansion was effectively at an end . . . when productions of space no longer pushed the borders of the unknown so much as re-worked its internal subdivisions. Faced with the loss of extensive nature, capital [has similarly] re-grouped to plumb an everyday more intensive nature."[4] This paradigmatic shift of focus from the expansive to the interior aspects of the natural world and from an extensive to an intensive project of exploration is nowhere more evident than in the emergence of contemporary biotechnology.

Over the past two decades, commentators from a range of academic disciplines have begun to speculate on the effect that the introduction of these new technologies could have on the ways that biological materials are understood, used, and valued. Such speculations began in the early 1980s as the first successful experiments in recombinant DNA engineering and gene transfer and synthesis were undertaken. Mooney and Doyle were among the first theorists to argue that biological materials take on a different status in the face of these new transformative biotechnologies.[5] They argued that where they had once been understood and valued simply as *material resources*, (sources of food, fiber, fuel, and so on) they ought now, as a consequence of this new-found ability to alter their fundamental genetic and biochemical properties and design, also be understood and valued as a type of *informational resource.*

The sociologist Manuel Castells made a similar argument in the late 1970s, when he began to write about the emergence of an information economy and the role of biotechnology within it. Castells, along with other social theorists such as Daniel Bell and Alain Touraine, had been predicting for some time, that capitalist development was about to undergo a major transformation: that much heavy manufacturing would decline and be replaced by a vibrant new information-based service economy.[6] This transformation would be induced, they suggested, by the development and introduction of a series of new, and linked, "informational technologies": electronic circuitry, microprocessors, and optical fibers. These technologies would provide new means of accelerating the transmission of information and, more significantly, new means of processing and reprocessing information in ways that could add value to it. Castells argued that biotechnology ought to be included as one of these informational technologies on the basis of its ability, as he put it, to "decode and reprogram the *information* embodied in living organisms."[7] New biotechnologies enable us, in other words, to extract genetic or biochemical information from living organisms, to process it in some way—by replicating, modifying, or transforming it—and, in so doing, to produce from it further combinations of that information that might themselves prove to be commodifiable and marketable.

The notion that biological materials contain valuable, accessible, and manipulable genetic information has solidified over time, now forming a central axiom of the life sciences. The distinguished historian of science Evelyn Fox Keller, commenting recently on advances in molecular and reproductive biology, argues, for example, that "the informational content of DNA remains essential—without it, development (life itself) cannot proceed."[8] The arguments put forward by Castells, Doyle, Krimsky, Keller, and others suggest that biological materials are currently undergoing another process of revaluation. No longer collected just for their material or corporeal properties (for their use as firewood or meat, for example) or for the insights that they might provide about other related species, they are now also valued for the "genetic information" embodied within them.

This argument seems entirely plausible; it is certainly seductive. The increasing use of metaphors such as "genetic information" and "genetic software" in both the popular press and academia are illustrative of that fact. Bio-informational metaphors are gaining currency everyday. Given this, it is surprising that relatively few attempts have yet been made to critically assess or refine the use of these bio-informational metaphors or to reflect on the complexities that attend their use.[9] It is clear that biotechnology has transformed the way in which biological materials are used and valued, but is important to be accurate both about the nature of these transformations and the terms in which they are described. The use of these terms raises important questions that warrant further attention. For example, what exactly do we mean when we say that biological or genetic materials are a type of information? Is this to say that genetic or biochemical matter is *analogous* to other types of information or that it is now *actually* a type of information? In other words, are such terms being used metaphorically or literally? If such terms are employed metaphorically, then what effect might this have on naturalizing the values or social or material practices that are implicit in the metaphor? If, conversely, they are used literally, what does this imply about the way in which biological materials are transformed by technological change? It's to a closer analysis of this questions that we now turn.

Genes as Information: A Cautionary Note

Perhaps it is inevitable, when seeking to describe a new phenomenon, to be drawn almost inexorably to the use of analogy and metaphor. As the geographer David Harvey has suggested, the deployment of metaphor lies at the root of the production of all knowledge, being "the primary means whereby the human imaginary gets mobilized to gain understandings of the natural and social worlds."[10] We can, he argues, "no more dispense with that imaginary than we could live without breathing."[11] The historians of science David Bohm and David Peet likewise suggest that metaphor is fundamental to science—both in extending existing thought processes and in penetrating "as yet unknown domains of reality *which are in some sense implicit in the metaphor*."[12] This raises though, key points: the relationship between metaphor and reality and the extent to which they remain distinct from one another. This question has taken on a particular significance in the technoscientific domain where the two appear to be collapsing into one other, making it difficult to establish a clear boundary between what is real and what is imagined. Metaphors, which are conventionally understood as discursive tools used to illustrate a certain process, are, in some instances it seems, now coming to function as literal descriptors of those same processes. Nowhere is this dilemma more evident than in the use of bio-informational metaphors—such as "genetic information" and "genetic software."

As the historians of science Donna Haraway, Lily Kay, and Emily Martin have all noted, uncritical adoption of the notion of "genes-as-information" can be problematic.[13] As Kay so tellingly revealed in her work "Who Wrote the Book of Life?", the history of the application of informational metaphors to biological function has a complex (and somewhat compromised) genealogy. As Kay reminds us, it is important to bear in mind that

> the information-based models, metaphors, linguistic and semiotic tools that were central to the formulation of the [notion of a] genetic code were transported into molecular biology from cybernetics, information theory, electronic computing and control and communication systems—projects that were deeply embedded within the military experiences of World War II and the Cold War.[14]

Before that time, as she points out, representational strategies in biology contained no references to information or language. However, with their importation into molecular biology, it became popular to conceive of organisms in this way—as complex command and control systems in which genes played a central role in both producing and regulating the circulation of genetic "information"—the blueprint, code, or message of DNA that would determine the form and function of the organism.

The application of informational metaphors within the biological realm has been viewed as contentious for several reasons. As both Kay and Haraway have suggested, there are considerable risks involved in retheorizing, and in the process reifying, living organisms as preprogrammed technological devices or mechanisms devoted solely to the production, transfer, and storage of information.[15] As both Kay and Haraway have argued, this type of representation is both reductive and essentialist. It invites researchers and the general public—to whom such concepts have been widely circulated—to imagine that complex processes of biological function and heredity are determined solely by a set of genetic "instructions"—a blueprint or code that alone provide the "essential" information for development—a kind of built-in program that directs and limits all possible biological outcomes. As Kay has so convincingly argued, the essentialist language that permeated information theory at that time—with its predilection for "frozen binaries" of the 0/1, yes/no variety, provides entirely insufficient tools for understanding or describing the complexity of biological function.[16] Divorced utterly from the nuances of environment and the possibility of chance alterations, biological study risked becoming "a kind of artificial life research, [in which] the paradigmatic habitat for life—the program—bore no necessary relationship to messy, thick organisms."[17]

Moreover, as both Kay and Haraway have argued, "informational" representations of life are also problematic in that they are invested with "extraordinarily potent *scriptural* images, tropes of *writing*, metaphors of language and the imagery of *texts* . . . the cultural potency of which has become doubly enhanced precisely because they [have come to] function *literally* rather than *figuratively*."[18] They argue that, as a consequence, this "book of life" is seen *not* as something that is actually being written or produced by scientists but rather something that exists independently of them, something *already written* (indeed, even preordained), there, waiting to be read, deciphered, and edited by "those who know."[19]

As Kay notes, it often assumed that informational metaphors were applied to the study of molecular biology because they were literally expressive of the inner logic of DNA, the actual architecture of the double helix. However, as she argues, these informational metaphors were imported into the discipline of molecular biology at this time *not*, as she argues "because they worked in the narrow epistemic sense (they did not) but because they positioned molecular biology within post-war discourse and culture, within the transition to a post-modern information-based society."[20] This is not surprising. As she notes, nearly every discipline in the natural and social sciences flirted with the seductive ideas of cybernetics and information theory in the postwar period. However, Kay agues that the embedding of the life sciences within a broader, emergent, information-processing revolution was particularly significant as it proved to have a profound naturalizing and, consequently, *operative* effect. Informational metaphors, which began as discursive tools for scientific or public persuasion (and which were reflective of a large-scale cultural shift in representation—the rise of an informational discourse) soon became literalized through a privileging of an informational approach to the study of molecular biology.

It is the operative effects of the naturalization of the informational metaphor that I wish to focus on here. Kay, Fox Keller, and others have revealed the impact that these informational discourses have had on directing the thoughts and actions of biologists and, hence, in shaping understandings of biological form and function. However, little attention has yet been given to exploring the effect and impact that an informational paradigm or worldview has had on the practice and politics of biological-resource exploitation. Before exploring these effects in detail, it is important to begin by understanding how and why this informational paradigm became so pervasive and dominant in the postwar years. This, in turn, requires us to understand something more about the broader changes that were occurring in the world's advanced societies and economies during these years—notably the rise of what is termed "the information economy."

THE RISE OF THE INFORMATION AND BIO-INFORMATION ECONOMIES

As I noted above, from the mid-1960s onward, social theorists such as Daniel Bell and Alain Touraine were beginning to predict the emergence of a new informational age.[21] The impetus for this development, they argued, would come primarily from transformations occurring in the economic sphere. The postwar era of production in manufacturing was dominated by a "Fordist" approach that had remained largely unchanged since the first moving assembly lines were put into operation at Ford's Highland Park, Michigan, plant in 1913 and 1914. The Fordist system of mass production was characterized by its rigidities: production was localized and reliant upon fixed plant and infrastructure, assembly-line manufacturing techniques, and set working relationships. As the heavy manufacturing sector began to collapse in the early 1970s, so too did confidence in this old regime of Fordist production. Viewed as increasingly fossilized and stagnant, Fordist modes of production were argued to have a constraining effect on capitalist development, one that might only be overcome by the introduction of more flexible and responsive forms of business organization and management.

This was reflected in the changing internal and external operations of firms. Rather than focusing on large-scale mass production, companies began instead to concentrate on specialization: low-volume production of a highly diversified product range and, most importantly, the continued research and development of new products. Production processes became more globalized as firms sought out lower-cost options, including cheaper labor and commodities. Companies either physically relocated production processes to developing countries or, alternatively, began to rely on international subcontracting. Strategic alliances, consortiums, joint ventures, and industrial cooperation and licensing agreements increased as companies sought new ways to extend production and consumption process beyond existing territorial borders.

Three facts became evident in this process of transformation. The first was that it would not be possible to co-ordinate the globalization of processes of production and consumption without profound improvements in the world's capacity for information transmission and processing. The second was that continued growth in the economy would increasingly rely on the research and development of new products. The third was that new industries could be created not only by buying and selling information but also by transforming it in ways that would add value to it. It was just at this time, in the decades that followed the Second World War, that a series of new technologies were beginning to emerge. These included the transistor in 1947, the integrated circuit in 1957, the planar process in 1959, the Internet in 1969, the microprocessor in 1971, and digital switching, the microcomputer, and optical fibers in the mid-1970s. The sociologist Manuel Castells has argued that they together combined to create a technological revolution—one dedicated specifically to this task of facilitating the transmission and processing of information.

It has been argued that all prior eras of rapid technological and economic change have been predicated on advances in information or knowledge generation and processing and that this "information revolution" was, therefore, no different in type from any that had occurred before. Castells argued however, that this new information age would be characterized not simply by the centrality of information (as this is indeed common to many phases of development), nor by the development of technologies that could facilitate the transmission of information, but rather by the creation of technologies able to act on or *creatively reprocess* information (by recombining, modifying, or replicating it) in ways that could add value to it. The significance of these new technologies, he argued, is that they have a *synergistic* effect. They act upon information in order to create from it other types of processed information that are themselves a source of productivity. It was predicted that with the development of these new technologies the world's economy would move into a new phase in which economic growth would rely not on industrial manufacturing but rather on the production, processing, distribution, and consumption of *information-based* goods and services.[22]

Within this context, it became evident that *information itself* would become an important industrial raw material. Particular types of privileged scientific or technical information could form the basis of new products, providing an important competitive advantage for those able to secure access to them. As Harvey has noted, in an informational society, "in a world of quick changing tastes and needs and flexible production systems . . . access to the latest technique, the latest product, the latest scientific discovery implies the possibility of seizing an important competitive advantage . . . knowledge itself becomes a key commodity to be produced and sold to the highest bidder."[23] Markets in information are, however, characterized by some rather unusual dynamics. As economists have famously noted, information is a unique commodity in that it is possible to give it away and yet still own it. This has important implications for the dynamics of exchange. When consumers buy stock resources, such as timber, iron ore, wool, or coal, they usually purchase them outright. "Ownership" in this context is absolute. This way of relating to property reflects an implicit understanding: these commodities will be consumed, and the consumer will, if necessary, return for further supplies, affording the seller another opportunity to create more income. Sellers accept that, at the moment of disposal, they relinquish their rights of ownership to the particular object or material of the transaction.

The producers of informational products (such as films or books) are not, however, governed by such constraints. They have recognized that they may exploit their assets without relinquishing complete control of them. They may decide to sell their products outright but may, alternatively, secure a further (and often greater) economic return by repeatedly selling *access* to the information embodied within them. Consumers of informational resources or products concur, signaling that they are less interested in owning the product than they are in securing access to the information within it, often on a short-term basis. For example, consumers of films and videos are often less interested in owning the actual film or video (in a material form — plastic case, film stock and all) than they are in securing access to the images and sounds contained within it. Those who enjoy books may wish to own the book itself, but may, alternatively, be happy just to read them. Equally, lovers of music may wish to buy recordings; on the other hand, they may be just as satisfied simply to listen to them.

This mysterious sleight of hand — the ability to access information without owning it and to give it away and still have it — favors particular types of commodity exchange such as rental and licensing. Repositories or storehouses of informational materials ("libraries" of books, videos, software, electronic music, or catalogued or archived data) may be accessed by those who wish to use the information without owning it, while those who control the repositories are able to secure an economic return each time the information is accessed and used.

Information is an interesting and unique commodity for another reason. The exchange value of information cannot be established with certainty until it is acted upon. Its future applications and hence future worth are unpredictable. This provides opportunities to speculate on its value, based on the various uses to which it might potentially be put. As Harvey suggests, all resources are products of "an ongoing technological, economic and cultural appraisal: to rent out a resource is also to rent out an imaginary — not only of what that resource is, but more importantly, of what it could be."[24] This is particularly so of information, which may be applied in many different ways and contexts. The same piece of information, while being of no use whatsoever to one party, may prove to be absolutely invaluable to another.

As the social anthropologist Marilyn Strathern has noted, assessing the value that attaches to information may thus prove difficult, as it may have to encompass "uses as yet unanticipated."[25] What is clear, however, is that real, rather than speculative, economic returns can be secured by convincing the market of the potential value of particular types of information and by charging consumers to gain access to that information—whether or not it later proves to have any actual utility for them.

An array of specialized consultancies and business-service providers have capitalized on this by speculating, dealing, and trading in raw information or by selling access to information-based products such as databases, computer software, and the like. If a fee is charged each time the information is accessed then the faster the information can be circulated and recirculated, the greater the rate of financial return. All forms of commodity exchange involve change of location and spatial movement and improving the efficiency of these movements has consequently been an important goal for all those involved in commercial processes of trade and exchange. Capitalism, as Harvey notes, has been characterized by continuous efforts to speed up the circulation of commodities and resources by overcoming the "friction of distance" that acts to impede their dissemination to producers and consumers. The distance involved may be small or large—it may entail speeding up the movement of raw materials from one country to another or may simply involve accelerating the rate at which they are cycled through the production process across the factory floor. One way of achieving this goal has been through the development of new technological innovations (such as assembly line production processes) that have enabled producers to dramatically shorten turnover times, thereby accelerating the rate at which new commodities are produced and consumed.

As I have illustrated here, from the mid 1960s onward a series of new technologies (computers, electronic banking systems, digital technologies) were introduced that were designed specifically to facilitate the transmission, storage, and reprocessing of information. These technologies began to have a dramatic effect on processes of trade and exchange, fundamentally altering the ways in which certain commodities were transacted. I would argue that there are important parallels between the way biotechnologies and other informational technologies act, but in order to draw those parallels out it might be helpful to analyze, in a little more detail, exactly how informational technologies work, what effect they have had on facilitating the transmission or processing of information, and how their introduction has changed the way existing commodity markets operate.

How Informational Technologies Act

In thinking about the way in which new informational technologies (such as computers) act, it is useful to return to Latour's analysis of the role that his "inscription devices" played in translating certain natural phenomena into more mobile, artifactual forms. He suggested that one of the greatest difficulties that colonial explorers faced was finding a way to bring home large, complex, or unwieldy natural phenomena without having to bring them home either for real, or for good. This dilemma was resolved by the development of new technologies (Wardian cases, nibbed pens, taxidermic tools, and so on) capable of translating natural phenomena into more transmissible forms—specimens, botanical illustrations, stuffed animals, and the like—that might effectively "stand in for" or "represent" that material in its absence. Something was necessarily lost in this process of translation, as certain properties were privileged and retained at the expense of others. For example, botanical illustrations convey nothing of the feel or smell of a plant; however, they provide other information about the plant (its color, shape, size, reproductive system) necessary for the purposes of classification. Inscription devices are paradoxical, then, in that they are designed to "retain simultaneously *as much, and as little, as possible.*" As Latour suggests, "this compromise between presence and absence is often called information . . . when you hold a piece of information *you have the form of something without the thing itself.*"[26]

Latour's analysis speaks directly to how contemporary informational technologies work. These technologies—which might be understood as archetypal inscription devices—operate on the same principle as their historical counterparts. They act on complex phenomena to create from them a series of new proxies that may serve as substitutes for the materials in question in their absence. Let me take, by way of example, books—or more specifically, a set of encyclopedias. Encyclopedias contain an enormous amount of information expressed as a text with images. This information is, however, embodied in a physical structure—a set of heavy books. The books have a particular materiality: they comprise velvety soft pages and hard covers; they are weighty, bulky, and have a lovely, musty smell.

With the advent of new informational technologies, it became possible to extract the information embodied in the encyclopedias (just the text and the images) and to render that information in an electronic or digital form as a computer data file, for example. Although it sounds obvious, even tautological, informational technologies act to privilege informational content at the expense of other attributes. It is evident, for example, that rendering the encyclopedias in an electronic or digital form entails a process of distillation — a kind of dematerialization.[27] The existing physical structure of the books is divested in order to purposefully retain and more effectively mobilize one "essential" component of it—the information contained in its text or images. Profound qualitative and experiential differences undoubtedly result, not least because something elemental, rather than just mass, has been lost in the process of translation. The feel, touch, and smell of the book are all sacrificed. However, this loss is accepted by both producers and consumers for one crucially important reason: the divestment of the physical structure of the book serves to facilitate the transmission of the information that was contained within it.

Similar processes are at work in other domains. The electronic symbols that flash across the world's financial markets now act as proxies for hard cash, while the digitalized music circulated by Napster acts as a proxy for other forms of recorded music that themselves act as surrogates for live performances. Books, music, and money can now be presented in new artifactual forms—as digital or electronic symbols or data, as information stored on databases, as MP3 files, or as electronic cash. There are, of course, obvious distinctions between the different modes. It is not the same to hear music on CD as it is to listen to it live; neither is it the same to read the newspaper online. Something is undoubtedly "lost in translation." However, while something has been lost, something else has been gained. Rather than being instantiated in its old materiality, the information in question finds itself housed in a new one and, consequently, amenable to a new way of life. Music, money, books, and so forth are all embodiments of information; these new technologies (or inscription devices) simply allow that information to be presented or rendered in new ways. Although this may not seem of any great significance, changing the way in which the information is embodied or presented proves to have profound effects on the dynamics of trade and exchange.

Proxies and Their Effects on the Social and Spatial Dynamics of Trade

The first effect of these transformations is that the information that consumers seek becomes available to them in variety of new forms. For example, the information in an encyclopedia becomes available as a book or as digital or electronic information stored on a database. Music becomes available as a live performance or as a CD or as a downloadable MP3 file. Money can be exchanged using banknotes and coins or through an electronic transaction. Dual or multiple markets emerge for the resource as it is variously constituted. Consumers are thus faced with a choice as to which medium they wish to use to access the same information. Proxies certainly offer a different and, arguably, more limited experience of the phenomena. It is equally evident, however, that consumers will accept proxies despite these experiential limitations. The question is, why do they? I want to argue here, that consumers accept and perhaps even prefer these new proxies because they offer them new and more efficient ways of collecting, concentrating and controlling, and recirculating and recombining to strategic advantage those "key" or "essential" elements of resources that they value most highly.

This is particularly so in circumstances where consumers do not value all attributes of a resource equally. Let me offer an example to illustrate my point. When undertaking research, I find that in many instances (although not all) what I most value in a book is the information in it—the text or images rather than the look, feel, or smell of it. For me, the information is the "key" or "essential" element of the whole—the physical structure of the book matters much less. I reveal the way in which I hierarchize the value that I attach to the different attributes of the book by signaling my willingness to accept a proxy that offers a different embodiment of the text and images—a photocopy or even a paperless, electronic version—one that offers me the text and images but not the book in which it was once embodied. This privileging process is, of course, entirely subjective. One makes assessments (both personal and commercial) of what attributes to retain or relinquish. A process of distillation or winnowing ensues. "Key" or "essential" elements are retained and privileged at the expense of other "unnecessary" elements, which are divested. This winnowing typically involves a shedding of "extraneous" or "redundant" material—a change in physical form, a literal "thinning out" of existing mass, body, or structure. The proxies that are generated out this process are, consequently, often more lightweight and transmissible than their existing counterparts, and consumers may come to favor them for this reason. They are simply easier to convey from one location to another; they provide a more *mobile and transmissible* embodiment of the material or information consumers seek.

The ability at least *to acquire quickly* the "necessary" or "essential" information that they seek (the text, for example rather than the pages that it is printed on) helps to override any remaining qualms that consumers may have about relinquishing a more complete experience of the phenomena in question (e.g., to be able only to read the information that was in the book rather than read the information while feeling the book, to be able only to hear the band rather than hear and see them.) Electronic publishing companies that specialize in converting books, documents, paper archives, drawings, microfiche, and microfilm into digital images using scanning, image formatting and indexing devices are now converting over 500,000 pages of text, pictures, or drawings from hard copy to electronic form *per day*. The artifacts or proxies that are produced from these processes—TIFF or PDF files—can be circulated and conveyed, that is, transported from one location to another, much more readily that the sets of books, archives, and libraries from which they were derived. It is similarly much easier to circulate money in an electronic form than it is to convey it as suitcases full of cash.

This ability to speed up the circulation of these commodities has, of course, had important economic implications. Books, data, financial resources, images, and so forth, which would once only have been accessible in particular locations, can now be accessed online or through other electronic media by a worldwide consumer base. Of course, access to these proxies remains socially and economically differentiated—not everyone has the necessary technology. However, for producers and consumers that have, these technologies provide new and infinitely more efficient ways of circulating or conveying information. This can make particular materials or information easier to access and collect.

Changing the way in which particular resources are embodied or presented can also serve to make them easier to concentrate, store, or archive. The information stored in whole libraries of books or music, and in archives of documentary records and the like, can now be compressed and re-presented as computer files, microfiche, or digital images. Quantities of information that could only have been stored or archived in vast spaces can now be stored in a single desk drawer. By using these new proxies it has become possible to concentrate huge collections of material and information in spaces and locations in which they could not otherwise have been accommodated.

New informational technologies, then, have provided new and much more efficient methods of transmitting and circulating information and of storing or archiving it for future use. However, as Castells noted, information technologies do not just speed up information exchange—they also provide new means of *acting on information* in order to create *new combinations of information* that may also be commercialized.[28] The proxies that are generated by information technologies are created by privileging the informational content of a range of different resources or materials at the expense of some, or all, of their other properties. Quite dissimilar materials can thus be rendered in a universal form—as a type of information, often digital or electronic. Rendering them in this way, in the form of a common "metalanguage," enables them to be read and manipulated by other information technologies. So, for example, scanning and computing technologies are used to convert the information and images in books into TIFF or PDF files.

When in these forms, information becomes more amenable to creative reprocessing using other informational technologies. Computers can be employed to combine and recombine, or substantively manipulate it to form new products (databases of new images, for example) that may be accessed electronically and searched interactively in ways that that would not have been possible had the information remained embedded in the book. MP3 technology has been used, similarly, to create downloadable computer sound files (MP3 files) from CDs that may be circulated via the Internet and then "sampled," modified, or recombined using other media technologies to create entirely new compositions and recordings. In both cases, the new combinations of information that are produced—the databases, images, or new musical compositions—are commodifable and marketable, providing important new sources of economic productivity for those who generate them.

The Generation of Bio-Informational Proxies

The introduction of these technologies in the postwar decades began to completely revolutionize approaches to information generation, processing, and handling. Together they combined to create a new economic paradigm—"informationalism"—and with it the rise of a new "information economy," one in which productivity derived from an ability to collect, concentrate and control, and creatively reprocess particular types of specialized information. An "informational" word view came to predominate. The life sciences were not, of course, insulated from these broader social and economic transformations, and it would be unrealistic to think that they did not begin to have an impact on thinking and practice in biology.

As Kay has suggested, although information theory itself had little application to biological research, the language that accompanied it—terms such as "code," "message," and "signal"—were eagerly adopted by life scientists, who conferred upon them a biological meaning even though, as she argues, they had none. Concepts and principles of information exchange, derived from the study of communication theory and cybernetics, came to inform analyses of complex biological phenomena. Protein synthesis, enzyme exchange, gene action, and, most particularly, heredity came to be understood as products of programmed communication systems governed by a genetic code—a notion that has undoubtedly served to essentialize and delimit understandings of the complexity of biological function and organization.

However, the impact that this wider informational paradigm and its attendant discourse has had on biology has not, in my view, been restricted solely to altering understandings and representations of the what might be termed the "microdynamics" of molecular-biological function—the inner workings of living organisms. I believe that it also had a profound effect on altering conceptions of how biological materials might be transferred or conveyed from one location to another—a broader dynamics of transmission and circulation.

"Information" is defined in the *Oxford English Dictionary* as "the action of telling or fact of being told of something," for example, news, intelligence, knowledge, or data. A time-honored but perhaps somewhat outmoded definition, "telling" in this sense perhaps implies that communication must necessarily be communicated face to face and orally. Without corrupting this definition unduly, it is possible, indeed appropriate, to give it a slightly more contemporary edge: to understand information as "something expressed or conveyed." As Kay has noted, informational metaphors were introduced to biology in the 1940s and 1950s when cybernetics and communications theory were revealing the many possible applications that command and control systems might have for guided weaponry, for thermodynamics, even, it seemed, for molecular-biological processes such as heredity. While this might explain why informational metaphors were first introduced to biology, it does not fully explain, for me at least, why they have continued to gain popularity long after cybernetics had gone out of vogue. This, I feel can only be fully explained by reference to the much wider information-processing revolution that began in a slightly later period, the 1960s and 1970s.

As I have illustrated here, it was during this period that the development of new informational technologies, such as television, computing and the Internet, was beginning to fundamentally alter conceptions of how information of many different types could be *expressed and conveyed*. By the 1980s, the potential applications and transformative effects of these new informational technologies were even more evident. Everyday processes of trade and exchange were being revolutionized by the introduction of electronic banking, personal computing, the Internet, and online marketing and through the creation and use of new informational proxies such as electronic data, digital images, compact discs, and electronic cash. There was undoubtedly a growing awareness that new ways were being found to present, circulate, store, and reprocess information that could provide important new sources of economic productivity.

It does not seem unreasonable to imagine that this new consciousness inevitably began to have an impact on approaches in the biological sciences. As the focus of biological research began to turn inwards to ever more detailed levels of molecular investigation, so emphasis began to shift away from the study of whole organisms and toward the examination and manipulation of the component parts of those organisms. The idea that these component parts of biological material could act as proxies for whole organisms in processes of research or development and/or yield up genetic or biochemical information that could form the basis of commodifiable products and processes began to animate entirely new approaches to the collection and utilization of biological material that have proven to have serious economic, cultural, and ethical consequences. It did so primarily by setting up a demand and an associated market, first, for these parts of organisms and, later, for other types of proxies—both corporeal and informational—that could be derived from them. It is crucially important to recognize how this market in new bio-informational proxies might operate, as I believe it serves to undermine existing and well-established markets for biological organisms (plants, animals, microorganisms, and so on) on which many depend for their livelihood.

The strategic advantages that accrue from an ability to "bring home" and study natural phenomena—particularly organisms—without having to bring them home either for real or for good have remained unchanged over time, and collectors and scientists have continually sought new means to effect this end. During the colonial era, the introduction of new technologies (such as Wardian cases, herbariums, specimen jars with screw-capped lids, and new preparations of preserving spirits) enabled the development of a series of artifacts such as, herbarium specimens and fixed-tissue samples that could act as proxies for whole organisms that proved too awkward or fragile to transport. While the project of translating whole organisms into new, more mobile artifactual forms remained of great importance, few technological advances were made in this regard in the first half of the twentieth century. However, from the 1950s onward, a series of technologies were developed that were to revolutionize completely the collection, transportation, and use of biological materials. These included portable cryogenic storage chambers, improved tissue-biopsy technologies, digital imaging technologies, and, most significantly of all, the introduction in the 1970s of equipment for extracting, amplifying, and sequencing DNA.

These biotechnologies had effects similar to the new informational technologies that were emerging at that time. Like informational technologies, they provided a means of acting upon complex phenomena (biology) in order to produce from it a series entirely new artifacts or proxies (cryogenically stored tissue samples, cell lines, and extracted DNA) capable of "standing in" for particular materials, resources, or phenomena (whole organisms) in their absence. Producing these artifacts or proxies also involves a very similar process of distillation. Much of the existing physical structure or body of the organism is divested in order to privilege and more effectively mobilize some other "key" or "essential" components of it—in this case the genetic and biochemical material embodied within the original organism. This process, which involves a progressive "decorporealization" of the organism, acts to facilitate the transmission of those key properties. The proxies or artifacts that are generated out of this process are, again, usually much more mobile (much more readily transported and circulated) than are the whole organisms from which they are derived.

It was possibly because such parallels could be drawn between the action of new informational technologies and biotechnologies that people began to characterize these "key" or "essential" elements as genetic or biochemical "information" rather than "material." It was at this time, for example, that Castells began to argue that biotechnologies ought to be included as informational technologies on the basis of their ability to decode and reprogram the "genetic information" embodied in living organisms. The use of such terms implies a degree of intellectual slippage—an inability to discern an apparently clear distinction between "things material" and "things informational." It also implies an inability to conscientiously patrol the boundary between metaphor and material reality—the boundary between what is real and what is imagined. However, as I suggested earlier, in the technoscientific domain at least, boundaries of many sorts—between "natural" and "artificial," "organism" and "machine," between "humans" and "animals," "material" and "information"—are blurring, making it difficult to discern where one ends and another begins.

Let me take these new proxies as a case in point. Most of the proxies that have been generated by new biotechnologies—cryogenically stored tissue samples, cell lines, even extracted DNA—retain some degree of corporeality; they remain, at some level a form of organic bodily matter. Others, such as sequenced DNA and MRI scans have none; all the body of the thing has been divested in the interest of rendering it in a more purely informational form—as data or image. If the presence or absence of corporeality were used as a criterion for determining whether a proxy is "material" or "informational" in nature, then it would seem that a clear distinction could be easily drawn. However, the process of "translating" biological material from a corporeal to an informational form is a progressive one, and determining the exact moment when "a body" becomes "a body of information" may prove difficult to discern. No better example of this can be provided than that offered by the case of the Visible Human Project.[29]

Begun in 1995 by the U.S. National Library of Medicine, and the U.S. National Institutes of Health, this project involved creating a digital archive of spectacular, three-dimensional images of the human body from two donated corpses—the first an executed prisoner and the second a Maryland housewife. Creating this comprehensive library of images involved an extraordinarily complex process of dissection and imaging. The bodies were first fully scanned using MRI technology. The cadavers were then cryogenically frozen to $-85°$ C in blue gelatin before being cut into four sections, each of which was CT and MRI scanned. These quartered sections were then planed at 1 millimeter intervals, using a highly sophisticated piece of technological "milling" equipment called a cryomacrotome, and then CT scanned once more. This complex technological labor generated the equivalent of 23 CD ROMs of digitized images of the ever-decreasing cadavers. The process of decorporealization was extreme, leaving nothing but a minute quantity of human sawdust. None of the corporeal mass of either body was retained—all of it was divested in the process of translating the bodies into archives of digitized information—three-dimensional interactive images—available for the purposes of diagnostic modeling, surgical planning, and visual pedagogy.

Similar processes have also been at work in the creation of genomic databases. With the introduction of DNA-sequencing technologies in the 1970s, it became possible to extract DNA from whole organisms and to elucidate and notate its structure—in other words to render it as a form of coded information, a literal text of Gs, Cs, As, and Ts that can be stored on large databases such as the Human Genome Project Database. This process involves an equally complex technical labor.[30] It is useful to know something about it for what it reveals about the blurring of the boundary between the corporeal and the informational.

Sanger's original technique for sequencing DNA involved copying DNA strands with dNTP (deoxyribonucleic acid triphosphates) and terminating them with tagged ddNTPs (dideoxyribonucleic acid triphosphates). These tags could be X-rayed to reveal the location of the nucleotides in the strands.

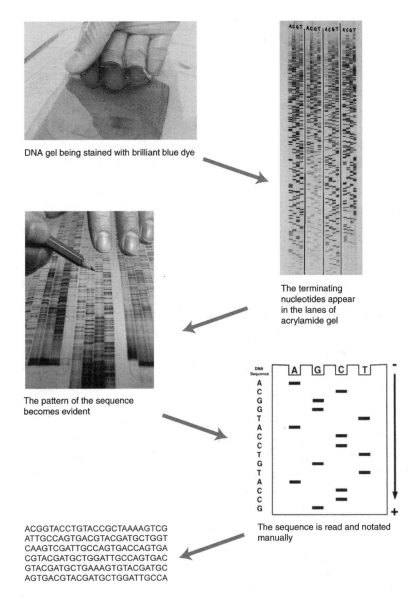

DNA gel being stained with brilliant blue dye

The terminating nucleotides appear in the lanes of acrylamide gel

The pattern of the sequence becomes evident

The sequence is read and notated manually

ACGGTACCTGTACCGCTAAAAGTCG
ATTGCCAGTGACGTACGATGCTGGT
CAAGTCGATTGCCAGTGACCAGTGA
CGTACGATGCTGGATTGCCAGTGAC
GTACGATGCTGAAAGTGTACGATGC
AGTGACGTACGATGCTGGATTGCCA

3.3 Reading bio-information: Establishing, noting, and notating the sequence of fluorescently stained terminating nucleotides in DNA gel. Images courtesy of U.S. National Institutes of Health and the Department of Genetics, University of Cambridge.

This technique was, however, very slow. During the 1980s, Craig Ventner began to develop technologies for automating sequencing operations. In automated sequencing, the nucleotides that terminate synthesis are labeled with different fluorescent dyes (G = black, A = green, T = red, C = purple). The colors distinguish which nucleotide terminates each cycle of DNA synthesis. A distinct banding pattern can be obtained when each of these fluorescently stained terminating nucleotides is separated out into a single "lane" of acrylamide gel. The positions of the nucleotides can then be read directly from patterns of the bands that appear in this gel (see figure 3.3) — either manually (by visual inspection) or automatically by a photocell that reads and records the sequence onto a computer database. As Hartl and Jones note, when used to maximal capacity, an automated sequencing instrument can ideally generate as much as 20 megabytes of nucleotide-sequence information per year.[31]

At this point the material exists in a curious kind of liminal state. When each of the flourescently stained terminating nucleotides is separated out into a single "lane" of gel, they can be read by the naked eye — the sequence can be determined by observation of the highlighted arrangement of the nucleotides. This engineered artifact is, simultaneously, corporeal and informational. It remains unquestionably a form of matter, and yet it also proffers itself as a source of immediately decipherable information that can be rendered in a textual form. It is suddenly not at all clear what the term "genetic information" means in this context. As Kay has noted, the term "information" and associated descriptors such as "code," "message," and "signal" have been used in wholly inappropriate ways to describe the actions of these nucleotides in "determining" form and function within organisms. However, it is also clear that the term "information" can also be used entirely appropriately, in my view, to describe the very real products of these other complex processes of material transformation. What begins as a thick, messy whole organism — something unquestionably corporeal in form — becomes, through this process, progressively decorporealized, existing in the final analysis as what could only be described as a body of information, an archive of stored data or images.

In both cases, biotechnology has enabled the genetic and biochemical resources embodied in whole organisms to be expressed or presented in a number of new ways: as cryogenically stored tissue samples, as cell lines, as extracted DNA, as digital images of tissue, and even as DNA-sequence information stored on databases. All provide important, commercially exploitable sources of what might be termed "bio-information." Changing the way in which this material or information is structured or presented may not, initially, seem a development of any great significance. However, I believe that trade in these new bio-informational proxies may act to fundamentally alter the existing dynamics of biological-resource exploitation. As I noted earlier, changing the way in which materials or information are embodied or presented can have profound operative effects on processes of trade and exchange. Just as new informational technologies have enabled particular resources to be presented in ways that have made them easier to circulate, store and reprocess, so biotechnology has enabled whole organisms to be rendered in ways that make the genetic and biochemical materials and information embodied within them much easier to transmit, store, reprocess, and recirculate.

This, in turn, has enabled collectors to speed up the social and spatial dynamics of collecting from which power and profit derive—the ability to acquire, concentrate and control, and then strategically recirculate collected materials to economic or political advantage. By concentrating on the acquisition of proxies, collectors are able to circumvent many of the complexities that are associated with the collection of whole organisms. Lucrative new industries are emerging that are devoted to the generation and marketing of these new proxies, industries that are populated by entrepreneurs who see an opportunity to capitalize on the particular qualities that these proxies offer—an ability to be readily transmitted, archived, reprocessed and recirculated. With this thought in mind, I want to turn now to consider how the creation of these new proxies might affect the social and spatial dynamics of collecting and, with it, the politics and practice of biological-resource exploitation.

Bio-Informational Proxies and Their Effect on the Social and Spatial Dynamics of Collecting

It could be argued that proxies such as fixed biological specimens, herbarium specimens, and anatomical and botanical drawings have existed for generations and that their existence has not yet served to radically alter the dynamics of biological-resource exploitation. One might wonder, then, why the development of new technologies and new proxies, such as cryogenically stored tissue, cell lines, or extracted DNA, would act to change these dynamics in any way. To understand, it is necessary to reflect again on the reasons people collect biological materials. As the earlier review established, motivations for acquiring biological materials are varied. People have long been interested in collecting biological materials for personal consumption, for pedagogical purposes, for display or classification, and as icons or exemplars. For these collectors, one individual specimen, either living or dead, might suffice to meet their needs. However, many other collectors have been motivated by a different rationale: the desire to acquire plants or animals, even human beings, with a view to transferring them to other locations where they could be successfully reproduced. Agriculturalists and horticulturalists were also interested in crossbreeding collected specimens in order to pass on particular genetically derived characteristics or traits such as color, variegation, pest resistance, and so forth.

This process of reproducing either whole organisms or specific traits could only be achieved through intergenerational breeding. Collectors first had to acquire and transport sound and compatible breeding populations to their preferred location, then maintain them in secure natural or artificial environments until they reached maturity or fruition, before overseeing and/or facilitating their reproduction through ongoing processes of procreation or germination. This often proved to be complex and time-consuming, as the history of the first attempts at transporting and reproducing rubber and breadfruit attest. Seeds proved to be useful proxies for whole plants—they were certainly easier to store and transport—however, they too had to be successfully germinated and raised to maturity and fruition. Many collected plants and animals were irrevocably damaged in transit or, alternatively, so radically stabilized to withstand transportation as to be rendered infertile. Proxies such as fixed biological specimens and anatomical and botanical drawings yielded information about the structure of the organism, but they did not provide a means of reproducing the organism or any part of that organism. Genetic materials remained embedded within whole organisms or seeds and could not be transported or reproduced independently of them. Anyone who wished to acquire, convey, reproduce, or modify genetic materials could only do so through the acquisition and breeding of these whole organisms or seeds.

These traditional dynamics of transmission and reproduction of genetic materials began to break down in the middle of the twentieth century as a series of new technological breakthroughs—the first ex vivo culturing of tobacco in 1944; the development of techniques for freezing bovine sperm in 1949, advances in in vitro fertilization, the introduction of recombinant DNA technologies in the 1970s, and advanced cloning techniques in the late 1980s and 1990s—allowed scientists to convey, reproduce, and engineer genetic materials independently of the organisms in which they were embodied. As research into genetic engineering and the synthesis and recombination of molecular compounds advanced, scientists began to recognize that they could accelerate the reproduction of genetic and biochemical materials (on which much of their research and experimentation relied) by simply isolating these materials from the organisms in which they existed and artificially replicating them in the laboratory under controlled conditions.

Where the transmission and reproduction of genetic materials had once demanded an "up-close" sexual interaction it could now be achieved by simply acquiring and replicating minute quantities of genetic and biochemical material. The reproduction of genetic materials no longer requires the amassing, in one location, of a viable fertile population; indeed, it no longer even relies on having possession of a single whole organism. Released from the biological, geographic, and temporal parameters that had traditionally governed the transmission of genetic and biochemical materials—the need for species compatibility, proximity, and fixed time scales of gestation—scientists (and the collectors that supplied them) began to turn their attention to new proxies that could provide them with just those essential or key elements that they needed—the genetic and biochemical materials that could act as starters for processes of artificial replication.

These did not include most of the proxies generated during the colonial era. They were not designed to preserve genetic material or information about genetic composition. In fact, the process of preserving biological specimens (whether through immersion in alcohol or through drying, treating, stuffing, or mounting) often had the effect of irrevocably damaging the cellular structure of the tissue, rendering them worthless as sources of reproducible genetic or biochemical material. However, new technologies such as portable cryogenic storage chambers, tissue-biopsy technologies, and DNA-extraction and -sequencing techniques could be applied to create a series of new proxies that retained viable and replicable quantities of genetic and bio-chemical material. Of course, not every researcher will be satisfied with a proxy. Those that still need to undertake work on whole organisms retain breeding populations. However, as the molecularization of biomedical and agricultural research has advanced, so the advantages of working with proxies, rather than whole organisms, become increasingly evident. Scientists were faced with a choice: they could continue to access the material or information that they were interested in through the acquisition and study of whole organisms, or they could alternately elect to seek out a proxy—a cryogenically stored tissue sample or cell line or even some genomic sequence data—that could prove *fungible*[32] for the whole organism (at least for the purposes of biomolecular research).

There are several reasons why scientists and collectors favor proxies. The first and most important of these is that they offer consumers a much more compact and readily transmissible embodiment of the material or information that they most seek. Bio-informational proxies—like other proxies—privilege and retain particular "key" or "essential" elements of the original (the genetic and biochemical material or information) at the expense of all other "inessential" properties (the majority of the physical mass or body of the organism), which are divested. The proxies that are generated out of this process are, consequently, much more "lightweight" and "mobile," much smaller, even infinitesimally smaller, than the whole organisms, and, hence, *much more transmissible*, easier to collect, transport, and circulate than are the whole organisms from which they are derived.

The second reason is that proxies afford scientists and collectors new and more efficient means of concentrating and storing genetic and biochemical material and information for future use. As I have illustrated here, in the past if scientists or collectors wished to maintain a supply of reproducible genetic or biochemical material, they necessarily had to obtain and sustain breeding populations in zoos, on farms, or in botanical gardens. That same material is now available to them in a more condensed form as a cell line or tissue sample. As a consequence, they may now store thousands and thousands of samples of reproducible genetic material from a host of different species in a space that is vastly smaller than that needed to maintain the same species as whole organisms. Just as processes of dematerialization have allowed vast quantities of textual information—the whole of the *Encyclopedia Britannica*, for example—to be reduced to strings of coded digital information able to be stored on one CD ROM, so processes of decorporealization allow large quantities of viable and reproducible genetic and biochemical material to be archived in one cryogenic freezer. By opting for the proxy, collectors are able to concentrate collected genetic and biochemical materials in specific centers or locations with far greater ease.

Changes in the spatial parameters of the traditional dynamics of the transmission of genetic and biochemical materials are not the only consequences of these developments; biological and temporal parameters are also being redrawn. Proxies are not only easier to convey and store than are whole organisms, they are also, thirdly, more combinable and manipulable. A rather different and more intimate dynamic of biological transmission is being altered as a consequence. Combining or manipulating different types of genetic material to produce desirable variations was a task that once required crossbreeding of existing species—a process that was governed, as I have noted, by certain biological, geographical, and temporal parameters. However, with the development of gene splicing, recombinant DNA techniques, and cloning, it became possible to employ proxies as substitutes for whole organisms in order to accelerate or even transgress existing processes of evolutionary development.

The procreative paradigm that has governed the nature and scale of transmission of genetic and biochemical material has now collapsed. It has become possible to combine genetic material from one organism with that of another, related or entirely unrelated organism without crossbreeding them. Hybrids have been overtaken by transgenic organisms and recombinantly engineered molecular compounds. There are now no "natural" limits to what Latour refers to as the "combinability" of these materials. The development, in California, of transgenic luminous lawns—designed to provide a form of night-time security lighting and created by splicing genes for luminosity found in deep-sea marine creatures into lawn seed—provides but one example of the potential for historically unprecedented cross-species integration.[33] These techniques have dramatically accelerated processes of manipulation and recombination of genetic traits that would either have been impossible to achieve through traditional breeding practices or would have taken generations to effect.

The economic advantages of accelerating processes of manipulation and recombination (on which most research and development in the life sciences are based) through the use of proxies rather than whole organisms are clearly evident, but they were well articulated in a recent article in *Scientific American*. Scientists who have been using clones while working on genetic modification in sheep recently reported that they now believe that it will be more efficient for them to simply inject DNA into cells, grow those cells in petri dishes, and then examine those cells for the desired genetic alteration. *Scientific American* was quick to point out the obvious economies of scale inherent in such practices, noting that "since it is so much easier to manipulate cells than sheep—not to mention the fact that it is easier to feed, say, 100,000 cells than the same number of livestock—much rarer and more subtle gene manipulation can be accomplished."[34]

Compounding the ramifications of these processes of decorporealization are others induced by the introduction of technologies such as cryogenic storage, which act to extend the temporal limits of the viability of these proxies. Cryogenic storage allows proxies such as cell lines, blood samples, and extracted DNA to be stored in a state of suspended animation for a potentially indefinite period of time, but certainly for at least several hundred years. Proxies that are collected now may also be archived and conveyed intact over both space and time for future reutilization. To avoid corruption or decay, they must be maintained in storage freezers, but this a task that certainly requires less labor and arguably fewer resources than that required to maintain breeding populations for the same periods of time. The combined effect of these processes has been dramatic: the amount of biological material required to constitute a viable industrial resource is growing ever smaller, while the same material remains available for use as an industrial resource for ever longer periods of time.

To summarize, scientists and collectors may favor proxies over whole organisms when undertaking processes of research and development for several reasons. Proxies privilege only certain key or essential elements of the original at the expense of other properties, which are divested. This process of decorporealization serves to make those elements (the genetic or biochemical materials embodied in the organisms) more accessible, more transmissible, and easier to collect, transport, store, concentrate, recombine, and recirculate.

The changes wrought by this process are even more pronounced when biological materials are rendered in wholly informational form—as a DNA sequence or tomographic scan. When rendered or expressed as electronic data, genetic and biochemical information becomes truly mobile, accessible from any portal across the globe. The data can be down- and uploaded in seconds, copied and circulated to hundreds of recipients instantaneously. At one time, information about the structure of an organism or its genetic composition could only have been derived from a firsthand analysis of living tissue, a task that again required the investigator to obtain whole organisms or samples of tissue. This information can now be acquired by simply analyzing genomic sequence data or images that can be downloaded and circulated from centralized databases. The transmission of genetic and biochemical information suddenly has the capacity to reach phenomenal speeds.

There are undoubtedly important continuities between these new bioinformational proxies and their historical counterparts. Illustrations, maps, sequences, scans, and images are all embodiments of information; however, altering the way in which that information is expressed can have a profound effect on the dynamics of use, trade, and exchange. When this information is rendered or presented in a digital or electronic form, it becomes amenable to the action of other information technologies. So, for example, when DNA sequence information is rendered in this way, it becomes possible for researchers working in different locations to access, circulate, copy, or store that information instantaneously. It also becomes possible for them to use that information in a more immediately interactive way. Researchers may, for example, search through vast quantities of information incredibly rapidly; they may download it, manipulate it using other computer software, or combine it with other downloaded sequence information in order to create entirely new databases of information. Digital images of human tissues can, similarly, be accessed, downloaded, manipulated, overlaid, and recombined to form new images that may then be circulated to other interested users. Each combination—each creative reprocessing of the material—may provide new and unanticipated sources of productivity—new commodifiable, information-based products.

When expressed in a digital or electronic form, *and only when in that form*, information derived from biological materials can be acted upon and interacted with in ways that would not otherwise be possible. As the sociologist of science Catherine Waldby noted when commenting on the Visible Human Project, "if human bodies can be rendered as compendia of data, information archives which can be stored, retrieved, networked, copied, transferred and re-written, they become permeable to other orders of information and liable to all the forms of circulation, dispersal, accumulation and transmission which characterize informational economies."[35] Of course, these particular proxies do not act as effective substitutes for whole organisms in all circumstances—however, they may in an increasing number of cases.

This has raised some concern about the effect that these bio-informational proxies might have on the existing dynamics of trade in biological materials. As John Barton, an international expert on intellectual property rights and genetic resources has argued, "the export (e.g. over the Internet) of a gene sequence from a nation is now the operational equivalent of the export of the organism containing the gene sequence. . . . As biotechnologists are increasingly likely to look to global genomic databases rather than to the underlying organisms from which the information is derived . . . genetic resource issues may soon be outflanked by genomic information issues."[36] As biological scientists have become aware of the tremendous functional utility of these new proxies, so demand for them has grown, providing the impetus for creation of new markets in "biological derivatives" of many different types.

By the 1980s, large corporations (particularly those operating in the pharmaceutical and agribusiness sectors) were coming to rely more and more on biotechnological techniques such as cloning and plant-cell cultures as a

Table 3.1
Plant products produced industrially using cell-culturing techniques by 1980

Plant	Secondary product	Currently produced in	Research by
Cacao	Cocoa butter (food)	West Africa, Latin America	Hershey (U.S.); Nestle (Swiss)
Chili pepper	Capsaican (flavor)	Latin America	University of Edinburgh/ Prutech (UK)
Cinchona	Quinine (drug)	Latin America, Asia	Plant Sciences Ltd. (UK)
Coffee	Coffee (food)	Asia, Africa, Latin America	Bio Foods, Fluor (U.S.)
Lithospermum	Shikonin (dye)	East Asia	Mitsui (Japan)
Pyrethrum	Pyrethrin	East Africa Latin America	University of Minnesota/Biotech (Belguim)
Saffron	Crocin (dye), Saffronin (flavor)	Asia	University of Edinburgh/Prutech (UK)
Sapota	Chicle (gum)	Central America	Lotte (Japan)
Thaumatococcus	Thaumatin (sweetener)	West Africa	Beatrice/Ingene (U.S.); Monsanto/DNAPlantTech (U.S.); Tate & Lyle (UK)
Vanilla	Vanilla (flavor)	Africa, Asia, Caribbean	University of Delaware/David Michael and Co. (U.S.)

Source: Kloppenburg, *First the Seed*, 275.

means of generating supplies of genetic or biochemical material. The first successful experiments in industrial-scale plant-cell culturing began in the 1950s and advanced rapidly throughout the late 1970s and 1980s. The technological improvements in cell culturing that arose from this experimentation during these years (most notably the development of large-scale industrial bioreactors) turned cell and tissue culturing from an unstable experimental procedure of uncertain outcome to an economically viable and sustainable method of producing secondary metabolites—the substances on which many pharmaceuticals, foodstuffs, insecticides, dyes, and other products are based.

The advantages of moving towards this means of production were clearly evident. All of the vagaries and uncertainties that imperil conventional agricultural production—environmental risks (drought, floods, and so on), political and labor instabilities in supplying countries (often in the Third World), uncontrollable variations in crop quality, crop adulteration, and losses in storage and handling—could be circumvented by bringing the production process under strict laboratory-controlled conditions. The comprehensive analysis of plant biotechnology undertaken by the rural sociologist

Jack Kloppenburg in the late 1980s revealed that interest and investment in in vitro reproduction of a wide range of important plant-derived commodities was already well advanced by the end of that decade (see table 3.1).[37]

Kloppenburg's foundational research, as well as work undertaken in the early 1990s by Goodman, Wilkinson, and Redclift, began to focus attention on the serious economic and ethical implications of using proxies (tissue samples or cell lines) as starters for the laboratory-based replication of biological commodities.[38] Many of these commodities have traditionally been grown in and exported from tropical developing countries. As these researchers noted, the one comparative advantage that these countries have always enjoyed in the production of these commodities has been their unique environment, which has produced the climatic conditions, soils, and complex ecosystems that could sustain particular species of plants and animals. While climatic conditions could be replicated artificially in other geographic locations using technologies such as glasshouses (at least on a limited scale), complex ecosystems could not. This advantage, they noted, would be entirely eroded by the development of technologies that enabled biological components to be artificially reproduced independently of the organisms, contexts, or locations in which they naturally occurred.

The implications of these developments were clear. Any acquired living sample of plant or animal material could potentially provide a means of replicating that resource in perpetuity. Moreover, these resources could now be reproduced in any location in which plant- and tissue-cell-culturing and, later, genetic-cloning facilities are found. There is undoubtedly a particular geography to the distribution of these facilities; they remain, like the greenhouses and other advanced technologies of the eighteenth century, concentrated in what are the contemporary equivalent of Latour's centers of calculation. Despite this, these facilities can *potentially* be reproduced in any location in the world, unlike the complex environments and ecosystems from which these biological materials are derived. As a consequence, it has been argued that the production of plants, animals, and their component parts has become *place nonspecific*. In commenting on these developments Goodman and Redclift argue that these facilities have the potential to "*trivialize* agricultural commodities" and to create new spatial divisions of labor by relegating developing countries to "the status of producers of low-value added biomass materials for generic, intensely competitive markets."[39]

Questions were raised about the impact that this transition to laboratory-based production could have on the economies of those developing countries that had once supplied agricultural and pharmaceutical corporations with raw materials. In proxies, corporations found a highly mobile embodiment of just those key or essential elements (genes, secondary metabolites, genomic information) that formed the basis of their products. By applying new biotechnological techniques to these proxies, they could potentially secure a much more controllable means of replicating the "raw materials" needed for production. The traditional dynamics of trade in biological materials—in which companies returned repeatedly to developing countries for ongoing supplies of raw material—were destined to be rewritten as a consequence.

These developments focused attention sharply on the question of the status and value of these proxies. Kloppenburg, Goodman and Redclift, and Buttel noted that the genetic and biochemical resources (both corporeal and informational) that were embodied in tissue samples and cell lines were, as a direct consequence of these developments, likely to become important commodities in their own right, the exploitation of which would provide important new sources of economic productivity in the burgeoning biotechnology industry. Mooney and Wilkes raised concerns about how bio-informational resources would be collected, used, and traded, noting that moves were afoot to introduce political and legislative mechanisms that would enable ownership of these bio-informational resources to be concentrated in the hands of First World corporations and research institutes.[40] There was certainly a growing awareness that while technological and economic change had provided the impetus for the creation of a new market in bio-information, it would be a third factor—regulation—that would shape the way in which this market would operate and who would profit most from it.

EMERGING MARKETS:
THE REGULATION OF TRADE IN BIO-INFORMATION

By the late 1970s, advances in biotechnology were fuelling an intense demand for genetic and biochemical resources, and trade in these materials began to burgeon. Despite this, no formal regulatory policies addressed the status of these materials or the terms and conditions under which they could lawfully be exchanged. Throughout the late 1980s and early 1990s, negotiations began to develop new protocols relating to the collection and use of these resources. These negotiations culminated in the ratification of two new sets of global regulations. The first of these was the Convention on Biological Diversity, ratified in June 1992, and the second was the General Agreement on Tariff and Trade's Trade Related Intellectual Property Rights legislation, or GATT TRIPs, ratified in 1994. These regulations had a particularly significant effect on trade in genetic and biochemical materials, as they constructed them (and/or modifications to them) as commodities to which private intellectual property rights could obtain. In other words, they acted to legitimate this trade by creating a space within which genetic and biochemical material could be exchanged on mutually agreed terms.

The question immediately arises as to why and how a set of regulations on intellectual property rights could come to pertain to biological material. Like the technological developments that had occurred in the 1970s, this regulatory shift had its genesis in the broader economic transformations that were occurring at that time. As I noted earlier, the economies of the world's most advanced nations were in crisis in the mid-1970s. Industrial manufacturing, which had driven economic growth during the postwar decades, had begun to collapse, triggering a review of existing economic and regulatory policy and a quest for new forms of economic productivity.

Three agencies had been responsible for directing and regulating economic development and reconstruction in the postwar years—the International Monetary Fund, the World Bank, and the GATT—or General Agreement on Tariffs and Trade. As its name suggests, the GATT was created in 1948, not as a formal agency, but as an interim or provisional "talking shop,"[41] a forum within which the rules for a prospective International Trade Organization might be negotiated. It was envisaged that the ITO would, in time, acquire a formal responsibility for setting worldwide trade regulations. The principles upon which this new regulatory framework was to be based were quickly established. There was a consensus of opinion among the architects of the agreement that the protectionist strategies of the 1930s had had a ruinous effect on the world's economy, contributing directly to the Great Depression. Their agenda was to reverse the effects of these policies by liberalizing trade relations. Although the proposed charter for the ITO was agreed upon in 1948, concerns in the U.S. Senate about the potential threat that a consensus-based model of diplomacy might pose to U.S. sovereignty acted to undermine commitment to the ratification of the organization. With ratification delayed indefinitely, the GATT remained the only instrument capable of setting world trading regulations. The GATT remained in place until 1994 when *The Agreement Establishing the World Trade Organization* was formally adopted.

Since its inception, the primary objective of the GATT has been to reduce or eliminate "barriers to free trade," such as tariffs or subsidies, which, it was argued, could be employed to shelter domestic manufacturing industries from international competition. This project proved so successful that the average tariff on manufactured goods fell from 40 percent to 5 percent during the total forty-seven-year period of GATT negotiations, during which worldwide trade volume increased fourfold as a direct consequence.[42] However, with a limit to further reductions on tariffs clearly in sight, the GATT began, at the commencement of its eighth round in 1986, to turn its attention to the deregulation of other *nontariff* barriers to trade. These were controversially argued to include national and international regulations governing local content laws, foreign investment, patenting rights, packaging requirements, and even occupational health and safety standards which, it was argued, could also be manipulated to shelter domestic industries from international corporate competition.

This argument provided the rationale for proposing that these disparate national and international regulations be "harmonized" under a new code of uniform global standards to be determined by the GATT. Following an intensive lobbying campaign, Europe, Japan, and the United States succeeded in having the GATT extend global-trade-related regulatory agreements to cover two new areas, one of which was Trade Related Intellectual Property Rights. With the introduction of this new measure, intellectual property, which had formerly been subject only to domestic or international regulation, became subject to new global-trade-related agreements.

Intellectual property refers to an invention, design, technology, or product created by a person or corporation. Intellectual property rights (such as patents, copyrights, or trademarks) are monopolies granted to inventors as a reward and incentive to creative activity. They afford the inventor exclusive rights of use, and the right to earn royalties from others' licensed use of their invention. Intellectual property rights are reciprocal—inventors are granted a monopoly over the invention for a specific period of time as recompense for their intellectual effort. In exchange, the inventor must agree to the public disclosure of the invention at the expiration of the term of the patent.[43] As information, ideas, inventions, and other forms of intellectual property have become increasingly important commodities in the global economy, individuals, corporations, and countries that are at the forefront of the development of informational products and processes have become anxious to ensure adequate mechanisms for their protection.

It was argued within the GATT that it was necessary to introduce a uniform code of trade-related intellectual property rights in order to prevent "unscrupulous" manufacturers, principally those in developing countries, from obtaining an unfair market advantage by engaging in the unauthorized reproduction of information-based products, processes, and technologies, such as computer software, microchip designs, electronic circuitry, and copyrighted brand names developed and patented in the First World. Establishing a *global* regulatory framework was deemed essential in light of the ease and speed with which these informational products and processes could be circulated, traded, and exchanged within the global economy. By sweeping away the host of different national regulations and instituting in their place a single regulatory framework, it became possible for the most powerful actors within the GATT (which included not only the G7 countries but arguably their most powerful transnational corporations) to be assured of global protection for the new information-based products and processes that they were rapidly developing.

The impulse to extend intellectual property rights to cover these new "inventions" was not, however, shared by all nations, and there proved to be little consensus within the membership of the GATT on the desirability of introducing such a broad protocol. While many member states supported its introduction, believing that it would finally provide a workable universal regime for prohibiting the unlicensed reproduction of commercially significant works and inventions, others objected, arguing that the protocol only served to protect the interests of developed countries who were already the world's largest producers and exporters of information-based products and processes. Many developing countries had operated either relatively informal systems of intellectual property rights law or none at all, believing that unrestricted access to new technologies and innovations would best serve to stimulate economic growth. The distribution and depth of coverage of Western systems of intellectual property rights was, therefore, geographically uneven.

Concern began to coalesce around the question of how the term "invention" would be interpreted: What would constitute an "invention"? Who would have rights to those "inventions" and on what terms? These issues were bought into sharp relief by the GATT's decision to include modifications to biological materials as "inventions" protected under the GATT TRIPs legislation. The decision to treat them as such had its precedent in judgments established under Euro-American systems of intellectual property rights law. The move to base a new and binding global regulatory regime on these selective judgments thus proved to be highly contentious, sparking a vitriolic debate about the legitimacy of the GATT TRIPS agreement.

The question of how biological materials came to be subject to domestic and later global systems of intellectual property rights is significant, not least because of what it reveals about the cultural embeddedness of law. For many, the law is seen as normative—expressive of a universally held and shared set of values to which all subscribe. However, on reflection it is evident that the law is a particular, culturally defined system for codifying knowledge. The judgments produced out of it inevitably reflect the concerns and interests of those who make them. When viewed in this way, law can be seen not simply as a set of received policies and practices but, rather, as a manifestation of ideology, "a system of signification which facilitates the pursuit of particular interests."[44] Examining the basis of justifications for extending patentability to modified biological materials may seem a tedious endeavor. In fact, it is actually a very important one, as the way in which decisions were made about what would and would not be patentable proved to have profound functional implications, determining which groups would be able to monopolize the use of collected genetic and biochemical resources and on what terms. In order to understand how these judgments were reached, it is helpful to begin with some understanding of the history and purpose of intellectual property rights law.

Intellectual Property Rights and the Regulation of "Embodied Inventions"

Proceeding from the assumption that every man (sic) already has "a Property in his own Person," John Locke argued in the eighteenth century that an individual's labor power also belongs to that individual. He concluded therefore that "whatsoever then, he removes out of the State that nature hath provided, and . . . hath mixed his Labour with, and joyned to it something that is his own, thereby makes it his Property."[45] Locke went on to argue that the right to a reward for expended labor should be enshrined in law, even in instances where the product of those labors was an immaterial thing such as an idea, invention, design, or literary work.

This argument met with considerable favor in Britain in the late seventeenth century. A primitive system of monopolies that afforded inventors the exclusive right to manufacture and use their inventions had first been employed in Italy in the thirteenth century to prevent the unregulated use of silk making devices, and this system was imported into Britain in the late sixteenth century. Queen Elizabeth I reportedly gave so many monopolies as favors to her friends that calls were made for a reformation of the system. The English Statute of Monopolies of 1642 created formal patent rights of fourteen years for "any manner of new manufacture."[46] The English statute established the reciprocal character of these rights.

Cornish has suggested that the statute served broader economic objectives—stimulating innovation, industrial growth, and employment rather than simply providing a "just" reward for expended labor.[47] The burgeoning of the Industrial Revolution, which was driven by the development of a host of new marketable commodities ranging from mechanical objects to literary works, many of which were based on protected innovations, ideas, inventions, and forms of codified information, suggests that this ambition was fulfilled. With the expansion of empire in the eighteenth and nineteenth centuries, Britain also discovered that substantial economic returns could be derived from the exportation of protected technologies and inventions. Patent applications increased accordingly, rising from less than 10 per year in 1750 to 110 in 1810 to 458 in 1840.[48]

America's Founding Fathers experienced an equivalent desire to ensure that American "authors and inventors" be able to secure limited but exclusive rights to "their respective Writings and Discoveries," and such rights were subsequently enshrined in the American Constitution. The ratification of America's first Patent Act in 1790 created an upsurge in patent claims, many of which were extremely dubious. The task of weeding out fraudulent and duplicate claims proved contentious and was not resolved in the United States until the Patent Act of 1836 introduced a requirement that patent applications be examined for novelty, utility, and patentability. Responsibility for adjudging what objects and phenomena might conceivably be disciplined or controlled within this particular regulatory system was vested initially with patent examiners but ultimately with the judiciary.

Intellectual property rights take a number of forms, including copyrights, patents, designs, and trade marks. Although there are important distinctions between these rights, they share a common rationale, providing protection for intangible things—ideas, inventions, symbols and information that have been manipulated or engineered through the application of human labor. As the legal scholars Bentley and Sherman have noted, in granting property status to intangibles the question arose as to how and where the boundary lines of intangible property rights were to be drawn. As they suggest, "while in real and personal property law questions of this nature are answered by reference to the boundary posts and physical markers of the objects in question, one of the defining features of intellectual property is that such markers do not exist. As a result, each area of intellectual property has been forced to develop its own techniques to define the parameters of the intangible property."[49]

Defining where this boundary between the "intangible" and the "tangible" lies has always proven to be problematic, not least because the idea, invention, information, or symbol to which rights are claimed must usually, in order to qualify for protection, find its expression in a material form. So, for example, in copyright law, the first general requirement for copyright to obtain is that that the work must be recorded in a material form—for example, it must be written down, taped, or filmed—"recorded in writing or otherwise." Writing is defined in the United Kingdom's 1988 Patent Act to include any form of notation or code "regardless of the method by which, or medium in or on which, it is recorded."[50]

Similarly, under patent law, ideas or information are not eligible for protection unless they have first been subject to some form of technical manipulation that involves reducing them to a concrete or physical form. Abstract, intellectual, and nontechnical phenomena including products of nature, discoveries, varieties of animals or plants, and natural biological processes for the production of plants and animals are not eligible for patent protection. Although the law does not explicitly characterize them as such, these abstract, intellectual, or nontechnical phenomena are understood as "foundational resources," part of a common heritage of mankind, upon which many different innovations and inventions might be based. A warrant for protection arises only in circumstances where the inventor employs his or her skill and ingenuity to reduce that abstract or intellectual concept or natural process to a specific, concrete, tangible, and technical, and thus patentable, product or process. Although what is actually being afforded protection is the author or inventor's intangible intellectual or creative labor, this labor must find its expression in a tangible object in order for it to be eligible for protection. The "object" or "work" exists as what might be called an "embodied invention": a tangible expression of an intangible or abstract creative thought process. This means, however, that it is not possible to protect this *intellectual labor* through the granting of monopoly rights, without also extinguishing others' rights to use *the object* or *invention* without license.

In order to be eligible for patent protection, an innovation must meet certain other criteria. First, it must be "novel." This is statutorily defined, in Euro-American law, as "non-obvious to a person skilled in the art." It must, second, involve an inventive step, third, have industrial application, and, fourth, be able to be adequately disclosed by description. While plant and animal breeders have always *theoretically* been able to apply for protection for their innovations under utility- or industrial-patent law, they have traditionally experienced difficulty in meeting the requirements for repeatability, inventiveness, and industrial applicability.[51] Objections to granting patent protection for plant varieties were also raised, initially at least, on the grounds that breeders were using "products of nature" and that these were ineligible for patent protection. Plant breeders were thus encouraged to develop separate systems of intellectual property rights protection such as Plant Breeders' Rights.

As new biotechnologies emerged in the 1970s, it became evident that modifications to biological materials would increasingly be effected not through traditional breeding practices but rather by engineering or recombining extracted elements of biological materials—genes, plasmids, biochemical substances, and the like. It was also clear that these modified genetic and biochemical materials might prove ineligible for protection on the basis that they a) were made from "products of nature" (genes, cells, tissues, and so on); b) were not necessarily "novel"; and c) could be construed as "discoveries" rather than "inventions." If these new products were to be made eligible for patent protection, a cohesive argument would have to be put forward that explained why these objects should not be excluded from protection on these grounds. The fact that most of the products and processes that were being generated in the life-sciences industry were being produced out of either very subtle manipulations of existing biological materials or through the acceleration of existing biological processes would make this task a very complex one. The economic advantages to be gained were, however, very significant indeed. A series of judgments made in U.S. legal jurisdictions proved to have far-reaching effects in this regard, providing the necessary rationale for extending patent protection to a range of previously unqualified biological materials. In the following section, I examine how these judgments were made, paying particular attention to the liberal interpretations that the judiciary had to make in order to extend the scope of patentability in this way.

Interpreting Requirements for Patentability

A long history of legal precedent in the United States had established that patent protection would not be available for innovations that were not novel, inventive, capable of industrial application, and able to be adequately disclosed by description. Abstract intellectual and nontechnical phenomena (such as theorems), discoveries, and varieties of plants and animals remained outside the scope of patentability as did products of nature and physical phenomena on the basis that they were considered "manifestations of nature, [that must remain] free to all men [sic] and reserved exclusively to none."[52] This latter exclusion was first seriously challenged in 1972, when a scientist working for General Electric, Ananda Chakrabarty, filed a patent application for a genetically engineered bacterium capable of breaking down the multiple components of crude oil. His patent claims included those for the process by which the bacterium was produced and for the bacterium itself as a human-made invention. The U.S. Patent Examiner allowed some claims but specifically rejected that for the bacterium on two grounds: on the basis that microorganisms are "products of nature" and that as living things they are not patentable subject matter under article 35, section 101 of the U.S. Constitution. Chakrabarty appealed the rejection of these claims to the Patent Office Board of Appeals, and the board affirmed the examiner on the second ground.

The decision was later appealed to the U.S. Supreme Court in *Diamond v. Chakrabarty* (1980), where it was overturned.[53] The bench ruled that although a natural material, the bacterium was not "a hitherto unknown natural phenomenon, but a non-naturally occurring 'manufacture' or 'composition of matter'—a product of human ingenuity."[54] Chakrabarty was deemed to have "invented" the microorganism by finding a way of successfully combining four plasmids and inserting them into a bacterium. In other words, he had, by the application of his technical labor and creative powers, induced a permanent physical change in some "natural" material in order to create a "non-naturally occurring" substance.

The decision to grant Chakrabarty the status of "inventor" proved to be contentious even within the judiciary. Four out of nine of the justices in the Court Of Appeal argued against the claim, noting that Congress, in drafting the patent legislation, had "chosen carefully limited language granting protection to some kinds of discoveries but explicitly excluding others, including living material."[55] The intention, they argued, was to ensure that individuals or industries were prevented from "securing a monopoly on living organisms, *no matter how produced or how used.*"[56] The decision also gave rise to a debate on the *degree* of inventiveness that must be involved in order to make good a claim for patent protection. While there is little doubt that Chakrabarty had used considerable ingenuity in creating the new organism, he himself remained somewhat dismissive of his inventive contribution. When asked to describe how he had produced the new bacterium, Chakrabarty reportedly replied: "I simply shuffled the genes—and changed bacteria that already existed."[57] His comments gave rise to serious concerns that monopoly rights would eventually be given for minor modifications to microorganisms, plants, or animals that, as Juma suggested, "may turn out to be no more than the embodiment of cosmetic improvements."[58]

Further concerns were raised by the judiciary's apparently inconsistent interpretation of "inventive contribution." Euro-American systems of intellectual property rights are deliberately selective in this regard, recognizing a right to reward for creative labor only in circumstances where the invention proves to be work of a single "author" and not where it proves to be the product of collective innovation. Recent work by anthropologists and critical legal theorists[59] has pointed out the troubling centrality that is given to the figure of the romantic individual author and his "original creations" in Western intellectual property law, illustrating the inordinate emphasis that it places on possessive individualism—what Gordon refers to as the "strong Lockean 'I made it—it's mine' justificatory pattern."[60] The difficulty is that the absence of an identifiable single "inventor" has often provided sufficient cause for applications for varieties of seeds and crops or medicinal remedies that are products of generations of *communal* innovation to be adjudged ineligible for patent protection. Although, they are also created (like Chakrabarty's bacterium) by modifying products of nature through the application of creative and technical labor, they are often judged to stand outside the scope of patentability. These interpretations have proven highly contentious, especially in cases where, for example, a grant of patent has been given in U.S. jurisdictions for a synthetically replicated medicament created in the developed world through the application of generations of applied indigenous knowledge. Although attempts have been made to include indigenous peoples as co-inventors on such applications, they have usually failed. A common argument for rejecting these applications has been that the information that indigenous groups provide about a plants location or uses is already in "the public domain" and cannot, therefore, be considered "novel."

Once again, much has come to rest on determinations of what is "novel" and what is not. If adjudged to be a novel innovation or novel application of an existing innovation, a grant of patent could be given; if a claim to novelty could not be sustained the application would fail. "Novelty" is statutorily defined in western systems of intellectual property rights law by reference to whether the invention in question would have been obvious to those "skilled in the art." In order to establish this, it is at first necessary to ascertain the "state of prior art" in a given field of endeavor. "Prior art" is interpreted very liberally in European and American patent acts, such that it includes any matter (whether a product, process, information about either, or anything else) that has been made available to the public (any public, anywhere in the world, no matter how remote) by written or oral description or in any other way. By working an innovation, describing it, or training another in its use, indigenous communities have, ironically, served to make their innovations part of the "prior art" and, consequently, ineligible for protection under Euro-American systems of intellectual property rights law.[61]

Modifications to plants and crops and varieties of seeds and production techniques developed for use in subsistence communities have also been excluded from patentability as they are intended for domestic, communal, or agrarian, rather than industrial use—a further requirement for patentability under Euro-American systems of intellectual property rights law. As opponents of biological patents such as Vandana Shiva have argued, this means that robust forms of property protection are available only to those inventors who innovate within an industrial mode of production. They are available, in other words, "only when knowledge and innovation generates profits and not when it meets social needs."[62]

Concerns were also raised about the legitimacy of legal interpretations of the distinction between "discovery" and "invention." As biotechnology advanced in the 1980s and 1990s, scientists were able to reveal the existence of a host of previously uncharacterized biological materials, including many genes and biochemical substances. Insofar as this research only served to uncover the existence of materials already found in nature, the question arose as to whether such endeavors should be understood as acts of "invention" or "discovery." While the Patent Act denies protection to discoveries, it limits that exclusion to "discoveries *as such.*" Interpretations within U.S. patent law have determined that an invention that makes *some practical use of a discovery* can be patented, as this is more than the discovery "as such." This practical use may involve simply isolating the material from its natural surroundings. As long as the material—such as a gene or biochemical substance—remains embedded within a whole organism it is not considered to be patentable; however, once these elements are extracted from the body by any technical process (no matter how straightforward), they become eligible for patent protection. Advocates of patenting argued that to treat the isolation of a gene or compound as a mere discovery would fail to provide sufficient reward for the work entailed in achieving this result. Opponents argued that to treat isolated genes or compounds as patentable inventions when the sole contribution of the "inventor" has been to reveal or discover an existing natural substance is to stretch the concept of "invention" beyond reason.

As Bentley and Sherman note, although the patenting of *processes* used to isolate natural substances is fairly uncontroversial, the same could not be said about the patenting of the substances that are isolated using these processes.[63] Serious questions have been raised about how it would be possible to "find" or reveal the existence of a substance embedded in a living organisms *without* isolating it from its surroundings. Although the patent is granted on the basis that the material has, through the act of isolation been transmuted from a "naturally" to a "nonnaturally" occurring phenomenon, there is nothing to suggest, as Sterrckx argues, that isolating a natural element from the body by technical means, or purifying that element, changes its "naturalness" in any way.[64]

Nevertheless, this rationale paved the way for patent rights to be extended to a series of elements derived from biological materials. Under previous forms of intellectual property rights, such as Plant Breeders' Rights, plants had always been treated as an indivisible whole. However, with decision handed down by the U.S. Board of Patent Appeals in the case *Ex Parte Hibberd* (1985), it became possible to patent not only multiple varieties of plants but also, "the individual components of these varieties: D.N.A. sequences, genes, cells, tissue culture, seeds and specific plant parts as well as the entire plant."[65] As Kloppenburg has suggested, the significance of the *Hibberd* decision was that it enabled "the licensing of particular *components*—e.g. a gene for herbicide tolerance—for use by a third party."[66]

Although individually these decisions seem relatively inconsequential, they together combined to enable patent rights to be extended to isolated genetic and biochemical materials and the engineered artifacts or proxies (genetic sequences, cell lines, and so on) generated from them, creating new levels of commercial protection and thus new incentives to trade in these materials. They did so by altering the way in which biological materials were understood—that is to say, "constructed"—and valued. Although these judgments can be seen retrospectively to rest on some rather selective and partial determinations (for example, what distinguishes an invention from a discovery, a "natural" from a "nonnatural" substance, a novel object or practice from a known one) they nevertheless served to create a new category of valorized objects (the biologically embodied invention) to which exclusive rights of ownership and use could obtain. In thinking about how processes of categorization can act to valorize particular materials and create new conditions for their use, it is possible to draw some parallels with another system for codifying knowledge about the natural world—Linnaean classification. Although the contemporary Euro-American system of intellectual property rights law and the Linnaean system of classification undoubtedly seem remote from one another, it is revealing to compare them for two reasons. First, I believe that the former is, in fact, predicated on the latter, and second, it has been through the geographic and cultural diffusion of these two mutually supportive, interrelated knowledge systems that the "disciplining" or regulation of much of the world's natural resources has, in different eras, been effected.

Coda: The Classification and Disciplining of Biological Material

As I argued earlier, biological materials were revalued during the Enlightenment once they had been ordered, that is to say, *rationalized*, within a particular, culturally determined system for codifying knowledge—the Linnaean system of classification. A value, both informational and material, accrued to those who were first to "extract" biological materials (whole organisms) from their "natural" surroundings (the wild) and to characterize them as classified "specimens" within this system. The criteria that were employed to create the fixed categories of classification were determined (in the Linnaean period, at least) in a somewhat arbitrary way. Nonetheless, the act of creating "standard" protocols for naming and characterizing materials had a normative effect—establishing apparently stable, universally agreed criteria for inclusion in the system, to which all contributors (collectors and scientists) could subscribe. The objective became clear: to "discover" "unknown" materials (unknown, that is, within this system) to classify and, in the process, characterize them as "specimens" (iconic exemplars that might stand for or represent whole "classes" of organisms) and then to locate them in their correct compartment within what might be visualized as "a grid of established relationships" (see figure 3.4). Much was to be gained from this process of "selective enframing." The materials came to acquire a further value as a consequence of being "known" within this system. A particular system of knowledge was built up about relationships between the collected organisms: it became easier to identify specimens of interest and their near relatives and to understand patterns of growth, inheritance, and acclimatization, knowledge, as I suggested earlier, that could expedite the further collection and utilization of particular biological materials.

Biological materials are currently undergoing another process of revaluation as a consequence of being subsumed within a new culturally determined system for codifying knowledge—the Euro-American system of intellectual property rights law. Once again, a value, both informational and material, accrues to the first party able to successfully place particular materials within a similar grid of established relationships. In this era, however, the aim is not to place existing "natural" organisms within the particular categories (class, order, genus, species, variety, and so on) that constitute the taxonomic framework, for this task has already (largely) been successfully

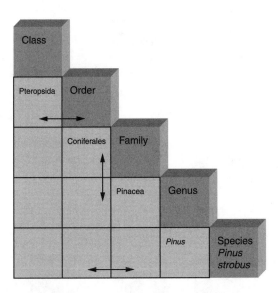

3.4 The Linnaean system of classification: A system of classi-fication based on a grid of established relationships. "With Lin-naeus's Tables we can refer to any fish, plant or mineral to its genus, and subsequently to its species, though none of us have seen it before. I think these tables so eminently useful that everyone ought to have them hanging up in his study, like maps" (Sir J. E. Smith, *A Selection of the Correspondence of Linnaeus, and Other Naturalists, From the Original Manuscripts*, 1821).

completed. Rather, the objective now is either to create "new" materials or to recharacterize existing ones such that they can be placed in the spaces *in between*—in what could be thought of as the *negative spaces* of the classifi-catory grid (see figure 3.5). A place within this new grid is an extremely valu-able thing to acquire, as it is rewarded with a patent: the exclusive right to use that material or resource for a given period of time.

Places within this new grid are reserved not for materials that have been classed into particular biological categories but rather for engineered bio-logical materials that are deemed to have *transgressed* the boundaries of these categories. Those organisms that already occupy a position within the taxonomic grid of established relationships—"known" species of plants or animals, for example—cannot be patented, as they are judged to be extant "products of nature." However, materials that are deemed to be "novel," that fall outside or between accepted categories of biological classification, may be valorized as "inventions." Just as a select group of eighteenth-

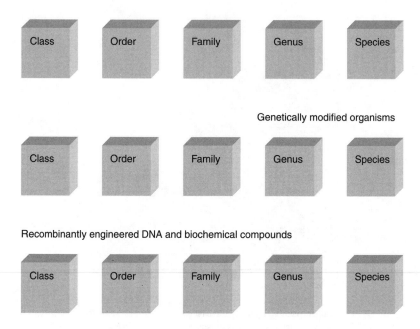

Genetically modified organisms

Recombinantly engineered DNA and biochemical compounds

3.5 The Euro-American system of intellectual property rights law: A system of classification based on a grid of nonestablished or novel relationships.

century European scientists set about establishing the criteria for inclusion in systems of biological classification, so a select group of twentieth-century Euro-American legal professionals (attorneys within the United Sates Patent Office and members of the U.S. Supreme Court) set about determining the criteria for inclusions in systems of intellectual property rights.[67] Although they too declared their practice to be rational and impartial, it is clear, in retrospect that this classificatory endeavor was, like Linnaeus's, based on some rather selective criteria.

Once again this raises a familiar question: of what possible relevance was this? Examiners within the U.S. Patent Office and patent law attorneys were, like their counterparts in earlier centers of calculation, busy creating a new cosmology—a new system for classifying and "disciplining" biological materials. By drawing on particular, culturally defined interpretations and determinations, this new system served to dramatically extend the scope of patentability to a host of previously ineligible objects and phenomena. This might have proven to be of limited significance had it not been for the fact that this system was quickly adopted as the basis, first, of other national systems of intellectual property rights law and, in time, of a new, global system of trade-related intellectual property rights law. In other words, these interpretations were initially confined to U.S. regulatory jurisdictions but were, as a consequence of these developments, quickly universalized to the global scale.[68]

It has been argued in recent times that the homogenizing force of globalization is so powerful that it has the capacity to create what Featherstone refers to as a "proto-universal culture."[69] While such terms are usually employed to refer to the globalization and homogenization of media and consumer culture, it is also possible to situate cultures of regulation within this context. As I have suggested here, classificatory and regulatory systems[70] can be seen as cultural constructs. These particular local and culturally determined systems for codifying knowledge can spread geographically, being adopted or imposed at a variety of scales. As I suggested in chapter 2, the Linnaean system of classification became internationalized in this way. It was however, a selective internationalization, as the system was adopted or imposed—that is to say, it colonized—along fairly narrow networks. Euro-American systems of intellectual property rights law have also spread geographically but, by virtue of the power of global institutions, have become much more pervasive, both geographically and culturally, in a much shorter space of time. With the ratification of the GATT TRIPs agreement, some 117 countries became *automatically required* to "make patents available for any inventions, whether products or processes, in all fields of technology, provided that they are new, involve an inventive step and are capable of industrial application" (article 27, 1).

Critics of the agreement such as Vandana Shiva[71] and Pat Roy Mooney[72] have argued that ostensibly "global" regulatory frameworks and policies such as those legislated under the GATT, although now universalized, are not normative but rather impose upon other countries and other cultures, "the globalized version of a very local and parochial tradition."[73] Shiva has argued that global regulations such as GATT TRIPs serve only the interests of First World elites and are in fact used as instruments to facilitate a process of "bio-imperialism" or "bio-colonialism."[74] She argues that they achieve this by requiring signatories to comply with the selective interpretations of "natural" and "nonnatural," "invention" and "discovery" laid down within this particular system for codifying knowledge (intellectual property rights law), a process that effectively demands that signatories recognize the right of First World elites to legitimately appropriate, privatize, and commodify all resources as yet "unclaimed" by this system. These concerns led to calls to abandon the TRIPs agreement altogether or, alternately, to generate an additional set of protocols that could ameliorate their potentially monopolistic effect. With this goal in mind, pressure mounted to include new regulations governing the use of genetic and biochemical resources under the Convention on Biological Diversity.

The Bio-Diversity Convention

The impetus to develop a strategy of global management for biological resource use came initially from conservationist NGOs such as the World Wildlife Fund and the International Union for the Conservation of Nature. These organizations had voiced early concerns about environmentally destructive practices such as overfishing and indiscriminate logging, calling for a coordinated international effort to stamp them out. It was evident, however, that countries already operated a multiplicity of (sometimes conflicting) national and international agreements relating to the use and conservation of biodiversity.[75] For regulations to be effective, it would be essential (as it had been with the GATT TRIPs agreement) that these different agreements be "harmonized" to form a new global protocol on the protection of biodiversity to which all nations could subscribe. The governing council of the United Nations Environment Program established a working party in 1988 to explore the possibility of implementing a new treaty on biodiversity use and preservation, reporting in 1990 that such a treaty was "urgently needed."[76]

Negotiations for the development of a new biodiversity convention began in 1991. Although initially designed to be a separate and autonomous agreement, the biodiversity convention soon became linked into the wider United Nations Conference on Environment and Development Agenda 21 declaration on environment and development. Although not negotiated directly under its auspices, the biodiversity convention undoubtedly reflects a dominant UNCED rationale apropos the use of biological resources. Introduced near the end of the Cold War as relations between East and West were beginning to thaw, the World Commission on Environment and Development sought to define global problems such as overpopulation, environmental degradation, food security, spiraling energy use, and urban decay as issues that could only successfully be resolved through close international cooperation and, most significantly, through continued economic growth. As its name suggests, the commission adopted an unashamedly utilitarian approach to the question of how best to protect biological diversity, linking environmental preservation directly to economic development.

The argument that runs throughout the commission's report is that living natural resources are best conserved both by and for economic development. Tropical forests, for example, are to be protected and conserved primarily because they are "crucial for development," acting as "reservoirs of biological diversity" that exist to be "developed economically."[77] As Middleton, O'Keefe, and Moyo suggest, wilderness becomes wonderful in this context "not just because of its fabulous beauty but also because it is seen as a repository of untold 'natural wealth.' "[78] The commission plays a crucial role here in embedding the notion that it is only by attaching a commercial value to wilderness that an incentive for its preservation can be found.

This "free-market environmentalism" gained momentum as formal negotiations for the UNCED conference began.[79] A number of commentators have suggested that the original agenda for the biodiversity convention—the preservation and conservation of biodiversity—was hijacked in 1991 at the point at which it became subsumed under the umbrella of the wider UNCED conference. As Chaterjee and Finger have argued, "concern for exponential destruction of the world's biodiversity [became] perverted into a pre-occupation with new scientific and biotechnological developments to boost economic growth."[80] The conservation of biological resources became inextricably linked to their "sustainable development." Biotechnology became constructed within the UNCED and later within the biodiversity convention as a useful tool for achieving these goals: providing new ways of exploiting biological resources and hence, new rationales for conserving them.

The drafters of the biodiversity convention initially adopted a very traditional "resource-management" approach to the use of biological materials, paying little attention to the impact of new biotechnologies or to the effect that they could have on way in which such resources could be used. It was not, for instance, until the G-77 group of developing countries demanded it that the issue of biotechnology and the use and control of genetic resources were included within the remit of the convention. Chaterjee and Finger have argued that the issue of ownership and control of genetic resources was not raised earlier by the developed countries, as "their major concern was protecting the pharmaceutical and emerging biotechnology industries, which get their raw materials from the forests."[81] Countries that had already been targeted by bioprospectors raised concerns that conventional resource-management strategies, while offering a means to protect the use of whole biological materials, would provide little in the way of protection for the use of the genetic or biochemical resources embodied within these materials. Conscious that these resources would form the basis of new, highly marketable products, they began to lobby intensively within the negotiating sessions for greater protection for these resources.

The conventions' final provisions on access and use of genetic resources are ultimately a compromise that seeks to balance the interests of both the suppliers and consumers of these resources. Significantly, the convention begins by abandoning the principle established by the Food and Agriculture Organization in 1983 that genetic materials constitute "a common heritage of mankind, [that] consequently should be available without restriction," preferring instead to construct them as potentially alienable commodities.[82] Interestingly, the convention employs two terms in the drafting: "genetic material," defined as "any material of plant, animal, microbial or other origin containing functional units of heredity"; and "genetic resources," which include "any genetic material of actual or potential value."[83] The convention goes on to introduce protocols that establish these genetic materials and resources as part of the patrimony of each nation state. States are at liberty to utilize them as they see fit, subject to certain guidelines.

It has been argued that the introduction of these provisions has acted to promote the further enclosure of nature by extending the commodity form and private rights of ownership to entities, such as genetic and biochemical materials and resources, that had not previously been subject to any form of direct regulation. An alternative view would suggest that these provisions have also provided a much-needed mechanism for enabling countries to secure a return for resources that would otherwise be appropriated without recompense. The rationale of the CBD clearly mirrors that of the UNCED conference in that it aims to achieve the conservation of biological diversity through the sustainable use of its components. The convention does, however, include an important and hitherto unprecedented caveat: it requires nation-states to ensure that the collection of use of genetic resources is carried out in a "just and equitable manner." It could be argued that the convention employs fairly unambitious notions of "justice and equity" in this regard; the wording does seem to suggest that both may be secured by simply rejigging the distributive paradigm such that the allocation of profit is more evenly apportioned. A less cynical reading would suggest that the measures reflect a new commitment to redressing past injustices and to ensuring that contemporary collecting programs are less exploitative than their forerunners.

Signatories are obliged, for example, under article 15 of the convention, to ensure that benefits arising out of the utilization of genetic resources be shared "fairly and equitably" with the suppliers of those resources, including indigenous communities that may have provided biological materials or information relating to their use. Under article 8, they are also obliged to regulate or manage the use of biological resources, including genetic resources, with a view to ensuring their conservation and sustainable use. These measures, when linked together, could, it is argued, serve a dual purpose. They could allow suppliers of genetic resources in developing countries to receive ongoing financial compensation for the successive uses that are made of collected genetic or biochemical resources and, in so doing, provide a further incentive to conserve the endangered ecosystems from which such resources are drawn.

If these measures prove to be effective, they may go a considerable way toward ameliorating the monopolistic effects of the GATT TRIPS legislation and fulfilling their much vaunted promise of creating a new, more equitable and just approach to the collection and utilization of biological resources. However, these goals will only successfully be met if the contracts that now govern the collection and use of genetic resources are capable of regulating what is, in effect, a global trade in biological derivatives—including types of bio-information. Recent experiences in other industries have revealed the many difficulties involved in tracking and monitoring the successive uses that are made of information-based products, such as a computer software, recorded music, or digitized images and data of many different types. Rendering these resources in a digitized or electronic form may have facilitated the ease with which they can be circulated around the market, but it has also made the task of tracking the circulation and monitoring the uses that are made of them infinitely more complex. A brief review of these findings may serve to illustrate some similar obstacles that will face those responsible for tracing and regulating the use of bio-informational resources.

Regulating "Spaces of Transmission"

While the task of controlling access to and use of "intangible property"— words, images, sounds, or expressed ideas—has always been complex, it has, as Pember suggests, been further complicated in circumstances where the work in question is widely circulated—for example, when it is published, transmitted, or broadcast.[84] A separate body of intellectual property rights law, copyright law, has been devised to address these particular issues. The principal aim of copyright law is to enable authors to protect their creative work from unlicensed reproduction by others. The fact that works such as maps, designs, musical scores, and fictional pieces were designed to be replicated and widely disseminated has always presented a challenge for those responsible for enforcing copyright but never more so than over the past fifty years, as new technological developments (the advent of photocopying, music and video recording, satellite communications, and cable television) have acted to create new and more extensive "spaces of transmission."

The task of monitoring performances of copyrighted musical works, for example, became infinitely more complex once they could be recorded and broadcast into thousands of different venues simultaneously. As Overbeck put it: "How could a composer or music publisher ever keep track of all the different radio stations, and night clubs, for instance, that are using his or her copyrighted material?"[85] In the United States, this problem has been solved by the creation of several music licensing organizations, such as the American Society of Composers, Authors and Publishers and Broadcast Music Inc., that represent the interests of composers, lyricists, and music publishers. These organizations use sampling techniques in an attempt to keep track of whose music is being broadcast where and to collect royalties on behalf of the copyright owners. Although these agencies have been relatively successful in their actions, there has, as Overbeck suggests, been "recurring disputes (and lawsuits) over the collection and disbursement of royalties for copyrighted music."[86]

The difficulties that regulators have experienced in tracking the unlicensed reproduction of copyrighted works have been greatly exacerbated by the development of new digital technologies. As Simon suggests, one of the effects of the digital age is to make physical copies of an author's work less valuable. As a consequence, "protection of the underlying intellectual property becomes even more crucial."[87] As a result of the development of new digitizing technologies, works that were traditionally distributed in print, disk, cassette, film, or other physical forms can now be transmitted in a digital form through electronic networks such as the Internet. As they can now exist as a sequence of binary numbers, it has become possible to create exact copies rather than close replicas of them. As media analyst Andy Johnson-Laird explained to those assembled at a meeting of the Electronic Information Law Institute in San Francisco in 1995, "digitized information can be copied quickly, easily and cheaply and the copy is quite literally every bit as good as the original. Add to that the notion that by standing on an electronic streetcorner, millions of passers-by can make identical copies for themselves in an instant and it is easy to see why the Internet, both physically and psychologically, is on a collision course with conventional copyright law."[88]

The infringement problems that arise as a consequence of these developments are myriad. The most significant involves the unlicensed copying of whole works, such as computer software programs, digitized images, music, or data created for use within computer networks. When in an electronic or digital form, works such as texts, maps and designs may be copied in seconds making the task of preventing unlicensed reproduction particularly difficult. These developments have provoked considerable concern amongst the leaders of America's software multinationals. The Business Software Association and its member companies, in their seventh annual study of global piracy of business software, conducted in 2002, estimated that 40 percent of all business software installed worldwide was unlicensed, a rise since 2000. As this figure suggests, despite the companies' concern and the investment of millions of dollars on the development of new sophisticated encryption and other access-control devices, there is little evidence of any slowdown in the unlicensed reproduction of these informational products.

Another problem relates to the unlicensed use of a *portion* of a work. Under the fair-use doctrine of copyright law, it is permissible to copy a small portion of a work without the author's permission. The copyright would, however, be infringed if an author's work or a portion of that work were incorporated into a new work without the author's permission. With the advent of new electronic and digital technologies, it has become much easier to copy portions of a work online (a sample of music from an MP3 file, a coded sequence of commands from a computer-game's program, or part or all of a digital image) and to then engineer them to form new works. Once the material has been modified, it becomes extremely difficult to establish its provenance, or indeed whether it had once formed part of a copyrighted work. There have been some successes in this regard—in the case of *Georgia Television Company v. Television News Clips of Atlanta*, for example.[89] The Atlanta company (a video-clipping service) was judged to have infringed upon Georgia Television's copyright by taking portions of their television news and public-affairs programs in order to create compilation videos of news clips regarding particular organizations. The prosecution was successful in this case, in part, because the portions, which were reutilized without permission, were still of an identifiable size and provenance, having been subject to relatively little modification.

While direct or literal copying is a very serious issue, it is still easier to regulate than what is called "substantial" copying. Taking caricatures of existing works as an example, Pember notes that in such cases "the defendant is not accused of taking a particular line or segment of a work but of appropriating *the fundamental essence or structure of that work.*"[90] For an infringement of copyright to have occurred, there must be more than minor similarities between the two works; they must be "substantially similar." If a work draws on the same basic elements of the original and employs them in a "similar" but not "substantially similar" design, no infringement has occurred. As Pember notes, however, "this is another one of the instances where it is easier to state a rule than to apply it—how can you determine when two works are "substantially similar?' "[91]

What lessons are there to be learned here for those who would seek to monitor the successive uses that will be made of bio-informational resources and prevent their unlicensed reproduction? New information technologies have induced profound changes in the way that certain existing resources can be rendered, and this has dramatically altered how they are accessed and used. Materials that would once have been conveyed manually from place to place can now be circulated instantaneously via the Internet, up- and downloaded in thousands of locations, copied, modified, recombined, and recirculated with ease. While these developments have facilitated the development of a new market economy in information-based products such as computer software, digital images and music, they have also served to make that market particularly difficult to regulate. The task of monitoring the flow of informational traffic and of detecting the unlicensed use, modification, or replication of information has proven to be immensely problematic.

Biotechnology, I would argue, has induced equally profound changes in the way that biological resources can be rendered, accessed, and utilized. They can now be conveyed in a partially decorporealized or even wholly informational form, as a sample of genetic material, a biochemical extract, or even as a coded sequence of DNA. When in these forms, they are much more transmissible, modifiable, and replicable. These factors have acted to create a new economy in bio-information, but they will also serve to make this new economy particularly difficult to regulate. As copyright lawyers such as Simon suggest, information technologies have transformed the dynamics of trade in copyrighted material in one crucially important respect: they have had the effect of making the physical form of a work less valuable while making the protection of the ideas or information on which the work is based increasingly more important. Or, to put it another way, they have revealed that what is now most valuable and requisite of most protection is not the *form* of a work but the *transmissible content* of that work.[92] With these considerations in mind, the question that remains is this: Will—or indeed can—the regulatory protocols devised under the biodiversity convention provide, as they purport to, an adequate mechanism for tracing and regulating the uses that are made of collected genetic and biochemical resources, even in instances where those resources are rendered in new, informational forms?

I have speculated here that a series of linked technological, economic, and regulatory changes have combined over the past two decades to transform the way in which biological materials are used as industrial commodities within the life-sciences industry. New proxies were and are being generated that can act as effective substitutes for whole organisms in processes of research and development. Where collectors and scientists might once have demanded whole organisms, they may now be satisfied with a tissue sample, a cell line, a biochemical compound, or even a sequence of DNA. These developments seem destined to transform the traditional dynamics of trade in biological material. Changing the way in which genetic and biochemical resources are embodied could, it seems, both enhance the collectors ability to concentrate, control, and recirculate them to their exclusive advantage, while simultaneously making the task of tracking and compensating for their use infinitely more difficult. What effect might this have on the practice and politics of biological-resource exploitation? Neither speculation nor theorization could, it seemed to me, provide an adequate answer to this question. That could only be arrived at through a detailed examination of the transformations that have occurred in recent years in the practice of collecting, storing, using, and trading biological materials as industrial commodities within the life-sciences industry. In July 1995, I set out to undertake just such an investigation, and in the following three chapters I report the findings of that research.

4. NEW COLLECTORS, NEW COLLECTIONS

The life-sciences industry, which can be broadly defined as including all those sectors that take the manufacture or engineering of life forms as the basis of their productivity, has grown dramatically over the last two decades as the biotechnological revolution has expanded the uses to which biology can be put. Industrial growth is often accompanied by a concomitant rise in demand for raw materials—in this case, for genetic and biochemical materials that might form the basis of a host of new products. As Sarah Laird and Kerry ten Kate estimate in their 1999 work, *The Commercial Uses of Biodiversity*, the market for products derived from these materials is an extremely lucrative one, estimated to be worth in excess of $500 billion annually.[1] A flourishing industry and market should make for a buoyant resource economy—suppliers of raw materials would ordinarily expect to see good returns under these economic conditions. But do they in this case? And if not, why?

These questions are not easily answered. In order to address them properly it is necessary to undertake a detailed analysis of why this market has emerged and of how it operates. It might seem too much to claim that this emerging trade in biological derivatives is in any way new or unprecedented. I believe, however, that the dynamics of collection and trade in biological materials have, in fact, been wholly transformed by the technological, economic, and regulatory changes that I outlined in the previous chapter. These have enabled genetic and biochemical resources to be extracted from the organisms in which they were previously embedded and constructed as alienable commodities to which private rights of ownership may obtain. These factors have combined, I suggest, to fundamentally alter approaches to the commercial and scientific use of biological materials—they have altered what is collected, how it is collected, and how those materials are stored, used, and later recirculated as commodities.

102

Complicating this further is that fact that these valuable resources may now be rendered in a variety of new artifactual forms—as cell lines, biochemical extracts, or extracted or sequenced DNA. These biological derivatives, whether corporeal or informational in form, provide important and highly transmissible sources of genetic and biochemical material and information, and they have consequently become highly valuable and highly tradable commodities. The collection of whole organisms and large tissue samples is clearly intimately linked to the production of these new technological artifacts, as the latter are created from them. But what effect might this capacity to render collected materials in new, more artifactual and transmissible forms have on the collector's ability to concentrate or store valuable genetic or biochemical resources or to transact them in novel ways? If a new trade in these biological derivatives and types of bio-information is emerging, who is it likely to benefit? Who will have the power to recirculate these new commodities, on what terms, and to whose advantage?

The agribusiness, pharmaceutical, horticultural, pharmacogenetic, and cosmetics industries are parts of the broader life-sciences industry, and each has been actively involved in collecting biological materials for many years. As Laird and ten Kate illustrate in their overview of the activities of these different industry sectors, each has organized its operations in slightly different ways. To reveal the nature of the changes that have occurred in collecting practice in recent years and their implications, it is useful to focus attention on just one of these sectors. I have concentrated here on the U.S. pharmaceutical industry, as they have one of the longest histories of involvement in natural-products research. In thinking about exactly how collecting practices have changed, I began by investigating how a combination of technological, political, legal, and economic events began to transform the way in which way biological materials were *collected and stored* for use as an industrial resource within this industry in the years from 1985 to 1995. In order to appreciate the nature and gravity of these changes, I begin by contextualizing them within a brief history of the early years of natural products research in the US from 1945 to 1985.

"WHEN THE WORLD WAS A KINDER AND GENTLER PLACE": EARLY PLAYERS AND VACATION PURSUITS

The project of collecting and examining biodiversity for pharmacological activity has a relatively long history in the United States. A range of publicly and privately funded organizations and institutions, including pharmaceuti-

cal companies, university departments, institutes of natural history, botanical gardens, and national research institutes have all played a role in collecting and exchanging materials gathered at home and abroad since the
since the early twentieth century. While many of these organizations were
primarily interested in collecting materials for study and pedagogical purposes, others, such as America's largest pharmaceutical companies, were interested in obtaining samples for applied commercial use. Although such
companies have always been interested in acquiring and examining exotic
specimens that might yield interesting biochemical compounds, they did
not begin to seriously invest in natural-products research until the 1940s.

The impetus for their involvement came initially from the state. The U.S.
government was, at that time, faced with a medical crisis, finding that they
had insufficient means to treat the growing number of war casualties.[2] In
seeking to improve methods of production of penicillin, they approached
Bristol Meyers Squibb, encouraging them to collect and test a wider variety
of soil samples obtained both domestically and internationally for more effective strains of mold. Other major pharmaceutical corporations such as
Merck Sharpe Dome, Smith Kline Beecham, and Pfizer also first became
involved in collecting and examining specimens obtained from abroad during these years. While interest centered initially on soil samples, natural-
products research later expanded, and the collection of higher plants, animals, marine organisms, and even insects became more commonplace.[3]
Any biological material that might provide a novel biochemical compound
was considered for investigation. Large scale investment in natural products
research was, however, mostly confined to the major pharmaceutical corporations. Although many antibiotics and over 25 percent of all prescription
drugs in the United States are derived from plant[4] or microbial materials,[5]
estimates that only 1 in 10,000 natural chemicals would produce viable
leads,[6] and that drugs would take an average of 12 years to develop[7] acted to
deter smaller companies. The costs and risks associated with undertaking
foreign collection programs were simply too great for them to entertain.

These corporate collecting activities were initially organized in an informal manner. Researchers working within Merck and Bristol Myers Squibb
would either undertake collecting expeditions themselves or "sporadically"
employ field researchers to collect biological materials on their behalf.[8]
They also relied on receiving materials donated from a variety of other
sources, including the occasional collector. As one senior researcher at Merck put it, "in the days when the world was a kinder and gentler place, samples were voluntarily sent in for screening by researchers and taxonomists

working in universities and the like. All sorts of employees were encouraged to collect things like soil samples while they were on vacation."[9] Other reports suggest that this practice was quite customary in the early years, with employees creating an informal society of collectors reminiscent of the Banksian network of dilettante collectors who sent material back to the centers of calculation in order to either curry favor or to advance the interests of science or commerce.

A number of universities and private research institutes in the United States also have long-standing interests in investigating the pharmacological properties of natural materials. Cornell University, the University of Chicago at Illinois, and the University of Mississippi, for example, have all undertaken long-term research into this subject, sponsoring collecting expeditions to many different parts of the world. Although their interest in pharmacology is principally academic, they have also contributed materials to industry for applied use. In the immediate postwar years, cooperation between university researchers and pharmaceutical companies was, as Dr. Borris at Merck has suggested, "pretty informal."[10] It was not uncommon to find university researchers and pharmaceutical companies routinely donating promising materials to one another for further investigation or commercial development.

Their experience of field collection was only rivaled by what might be thought of the "New World centers of calculation": North America's largest museums of natural history and botanical gardens. The New York Botanical Garden, the Missouri Botanical Garden, the American Museum of Natural History, and the Smithsonian have all historically sought to procure exotic specimens for research or display, and most have been actively involved in collecting for over a hundred years. These institutions are also repositories for "orphaned" collections of material vested to their care. But perhaps the largest collections of biological materials that have been created in the United States have been amassed over the last fifty years by two state-funded agencies, the U.S. Department of Agriculture and the National Institutes of Health. Both organizations have sponsored many international collection programs. The fungal and microbial material, herbarium specimens, and crop germplasm collected by the USDA constitutes the largest collection of archived biological material in the United States. The principal focus of USDA research has been the development of new cultivars; however, material from the collection has also been used for pharmacological purposes—the penicillin that the U.S. government developed during the Second World War was created from fungal material stored in

USDA repositories—and their collection remains of continuing interest to the pharmaceutical industry.[11]

The NIH, the largest government-funded health-research institute in the United States has, conversely, an *explicit* obligation to research and develop new therapeutic treatments.[12] Natural-products research has long been considered a risky venture, one most likely to be undertaken by those who need to develop novel or radical approaches to the treatment of disease. This may perhaps explain why the National Cancer Institute has the longest continuous involvement in natural-products research of any government-funded agency in the United States.[13] In order to protect their nonprofit status, government-funded organizations such as the National Institutes of Health may only advance research and development work to the stages of preclinical and clinical trials, after which point they are required by federal regulations to license out commercial production of any drug to a registered pharmaceutical company. In the interests of furthering research, the NIH may, however, accept for testing purified natural products that have either been submitted voluntarily or solicited from other sources.[14]

The National Cancer Institute instituted its first, formally organized, large-scale natural products collection and screening program in 1960.[15] Unlike earlier efforts, which were organized on a fairly ad hoc basis, this new collection program was undertaken systematically and in association with the USDA. By drawing upon the USDA's collection agencies, the NCI was able to secure some 35,000 samples of plant material, 500 different species of marine organisms, and over 180,000 microbial extracts from North America, Mexico, Europe, and Ethiopia for pharmacological screening. Although two important new drugs, Taxol (paclitaxel) and camptothesin, were successfully produced from these materials during this period, the program was scaled down in 1980 and abandoned altogether in 1983 as the researchers felt that the collected materials had been examined as thoroughly as possible and were unlikely to yield anything new. The remaining samples of material collected under this program were destroyed to free valuable working space. As the director of the Natural Products Branch of the NCI, Gordon Cragg, has suggested, the NCI's decision to discontinue their collection program had an adverse affect on natural-products research generally and particularly within the largest pharmaceutical corporations, simply because "the way NCI goes in this country . . . tends to dictate the way interest in natural products goes. It is an inevitable consequence of the fact that we are a major organization that has the largest program in natural-products research."[16]

Natural-products research did, in fact, go into a decline in the United States from the late 1970s onward, particularly once word leaked out that the NCI was intending to abandon its foreign collecting operations. The Rural Advancement Foundation (a U.S.–based nonprofit research institute) has suggested that by 1980 very little of the U.S. pharmaceutical industry's research budget was being spent on natural-products research.[17] With the exception of the collection programs that were undertaken to support the pure research efforts of scientists and academics working in university departments and research centers, few other biological collection programs were undertaken by or on behalf of the U.S. pharmaceutical industry during the late 1970s and early 1980s.

"AN HISTORIC REVIVAL OF COLLECTING"

1985 marked the commencement of what has been described elsewhere as "an historic revival of collecting."[18] This revival, which was evidenced by the establishment of a multiplicity of new programs, although still continuing, appears to have peaked in the decade from 1985 to 1995. Although relatively little time had elapsed since the NCI's last major collection program had ended, considerable changes were made in the way these new collection programs were organized—these changes were significant enough to distinguish this new era of collecting from any that had preceded it. The first major collection program to be undertaken during this period was launched by the NIH in September 1986. This program, which was to run for five years, seemed at first to be identical to those undertaken by the NIH in the years from 1960 to 1980: it would be undertaken collaboratively with a view to securing large numbers of biological samples from a wide geographic area. The nature of this collaboration would, however, be fundamentally different to that which had existed before.

In 1985, the NIH decided that it would be profitable for them to broaden the number and type of organizations and institutions that collected on their behalf. Following a process of formal tendering, collecting contracts were offered to three organizations: the Missouri Botanical Garden, the New York Botanical Garden, and the University of Illinois at Chicago (assisted by the Arnold Arboretum and the Field Museum in Chicago).[19] By engaging the services of these organizations, the NCI could extend the geographic scope of the collecting operations to include biodiverse regions that it had been unable to explore with its former partner, the USDA. As the head of the NCI's Natural Products Branch explained,

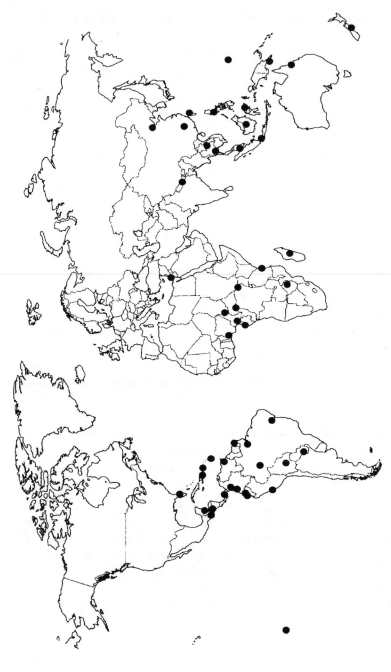

4.1 Map of the NCI's collection sites, 1985–1995.

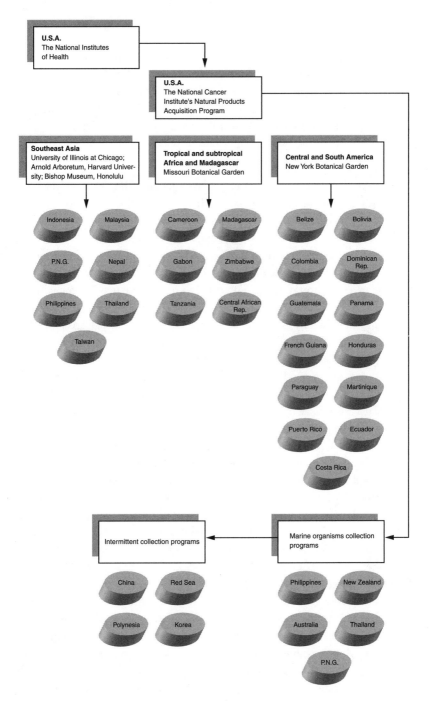

4.2 Schematic diagram of the NCI's hierarchical network of collection, 1985–1995.

We had an interest in going to the tropics because we hadn't been there to date and we knew that organizations like New York and the Missouri Botanical Gardens were pretty active in South America, Africa, and so forth. And the USDA does have wide-ranging programs, but they didn't have the particular focus on tropical countries that these other organizations had, and so we figured, let's go outside, to the private sector.[20]

The NCI contracted these intermediary agencies (the U.S.–based botanical gardens and museums of natural history) and they, in turn, subcontracted similar institutions in developing tropical countries to carry out the collections in the field. As we shall see, these "indigenous" institutions were perceived domestically to represent local interests, and they played an more important role in both facilitating and, perhaps more importantly, legitimating the collection process. With the signing of these new contracts, a formal subcontractual relationship was set up between a master-collector (the NCI) and a subhierarchy of collecting agents. The contracts were subject to competitive renewal in 1990 and were then extended for a further five years. Through this system of subcontracting, the NCI was able to create a network of collectors and a geography of collection, the scale and extent of which was astonishingly wide and certainly unprecedented in the area of pharmacological research (see figure 4.1, figure 4.2, and table 4.1). Collections were undertaken in over forty countries during these years by employing the services of some forty-five local collection institutes.[21] The first truly globalized, US based, biological collection program had been created.

Just as the NCI's decision to terminate their collecting program in 1980 had initiated a crisis of confidence in natural-products research, their decision to undertake this expansive new collection program was to have a generally stimulative effect. This was reflected in the sudden rise in the number of major pharmaceutical companies that began to institute new collecting programs from 1987 onward. As Reid et al. have documented, Smith Kline Beecham, Upjohn and Co., Monsanto, Ciba Ceigy, and Boehringher-Ingelheim, among many others, all began to institute biological collection programs at this time, employing intermediary agencies such as universities or botanical gardens to undertake the work on their behalf. In September of 1991, in what was hailed as a groundbreaking agreement, Merck announced that it was entering into a new bioprospecting partnership with Costa Rica's National Biodiversity Institute (InBio). Under the terms of the contract, InBio agreed to provide Merck with samples of plants, insects, and

Table 4.1 The NCI's network of collecting agents, 1985–1995

Collection site	Contracted collecting organization	Collected materials
Australia	Australian Institute of Marine Science, Townsville, Queensland	Marine macroorganisms
Belize	Ix Chel Tropical Research Centre, San Ignacio	Plants
Bolivia	National Herbarium, La Paz; Museo de Historia Natural, Santa Cruz.	Plants, microbial materials
Cameroon	Institute for Medical Research and Studies of Medicinal Plants, Yaounde	Plants, fungi
Central African Republic	Not available	Not available
Central and S. America	New York Botanical Garden	Plants, fungi, microbial materials
China	Kuming Institute of Botany	Plants
Colombia	Universidad de Antioquia, Medellin; Universidad del Valle, Cali; Jardin Botanico, Tulua	Plants, fungi
Dominican Republic	Herbario "Dr. Rafael N. Moscoso," Santo Domingo	Plants
Ecuador	Fundacion Natura; Pontificia Universidad Catolica del Ecuador, Quito	Plants, fungi
Florida	Harbor Branch Oceanographic Institute	Marine macroorganisms
French Guiana	Not available	Not available
Gabon	Herbier National, Libreville	Plants
Guatemala	Universidad de San Carlos, Guatemala City	Plants
Honduras	Fundacion Hondurena de Investigciones Agricolas	Plants, microbial materials
Indonesia	Herbarium Bogoriense; Council for Sciences of Indonesia	Plants, fungi
Madagascar	Centre National de Recherchés Pharmaceutiques, Antananarivo	Plants, microbial materials
Malaysia	Department of Forests, Sarawak; Institute for Advanced Studies, University of Malaysia, Forest Research Institute of Malaysia, Kuala Lumpur	Plants, fungi
Martinique	Galerie de Botanique, Fort-de-France	Plants
Nepal	Department of Forestry and Plant Research, Kathmandu	Plants, microbial material
New Zealand	University of Canterbury, Christchurch	Marine macroorganisms
Palau	Coral Reef Foundation	Marine macroorganisms
Panama	Not available	Not available
Papua New Guinea	Forest Research Institute Lae, Lae Herbarium; Department of the Environment and Conservation, Boroko	Plants; marine macroorganisms
Paraguay	Facultad de Ciencias Quimicas, Universidad de Ascension	Fungal material, microbial material
Peru	Instituto de Investigaciones de la Amazonia Peruana	Plants, fungi

(*continued*)

Table 4.1 The NCI's network of collecting agents, 1985–1995 (*continued*)

Collection site	Contracted collecting organization	Collected materials
Puerto Rico	Not available	Not available
Philippines	Philippines National Museum, Manila; Forest Products Research and Development Institute, Los Banos; organizations co-coordinated by Dr. Ernai Menez, director of the Smithsonian Oceanographic Sorting Centre	Plants; fungi; marine macroorganisms
Polynesia	Brigham Young University	Plants, fungi
Red Sea	Tel Aviv University	Marine invertebrates
Southeast Asia	University of Illinois at Chicago, Arnold Arboretum, Harvard University, Bishop Museum, Honolulu	Plants, fungi, microbial materials
Taiwan	Chinese Culture University, Taiwan Forestry Research Institute, Taiwan National University (all in Taipei); Taichung Medical College; Hengchun Tropical Botanical Garden	Plants, fungi, microbial material
Tanzania	Department of Botany and Traditional Medicine Research Unit, University of Dar El Salaam	Plants, fungi
Thailand	Royal Forest Herbarium, Bangkok; Institute of Marine Science, Bangsaen	Plants; marine macro-organisms
Tropical and subtropical Africa and Madagascar	Missouri Botanical Garden	Plants, fungi, microbial materials
Zimbabwe	Zimbabwe National Traditional Healers Association	Plants, fungi

Source: Collated from information provided from the Natural Products Acquisition Program of the U.S. National Cancer Institute.

microorganisms from Costa Rica's conservation areas in return for a two year screening and research budget of $1,135,000 and an agreed royalty on the sale of any products developed from these materials. InBio agreed to contribute 10 percent of the budget and 50 percent of any royalties to Costa Rica's National Park Conservation Fund. In addition, Merck also agreed to provide technical assistance and training to improve Costa Rica's drug-research capability.[22] The agreement reflected a serious formalization of the company's approach to biological collecting. For Merck, the days of amateur, vacation collecting were over.

Merck's announcement stimulated a great deal of public and corporate interest in bioprospecting. The fillip given to natural-products research by the introduction of this bioprospecting program was reflected in the decision-making strategies of other pharmaceutical corporations. Sterling Winthrop and Bristol Myers Squibb both let new collection contracts in the early 1990s, as did Pfizer, who announced their intention to undertake a joint bioprospecting project with the NYBG in 1993. Attracting less attention but equally as significant were those collecting programs initiated by smaller companies. California's Shaman Pharmaceuticals, which was established in May 1989 with the exclusive aim of "discovering and developing new classes of pharmaceuticals derived from tropical plants,"[23] launched the first of several bioprospecting expeditions to Peru and Ecuador in 1990.[24] A host of other small companies that specialize in the collection and supply of biological materials for use in the biotechnology industry, such as Pharmacognetics of Maryland[25] and Phytera of Massachusetts,[26] also stepped up their collection activities from 1990 onward, establishing new links with academic and scientific collecting institutes.

In addition to this, a number of even smaller and more heterodox companies began to express an interest in bioprospecting after 1993. The Knowledge Recovery Foundation, for example, which operates out of the second floor of a brownstone in central Manhattan, launched a fund-raising initiative in 1995 to underwrite the cost of establishing, in New York, a "library of medicinal plant material and information" that the organization had begun to collect in "indigenous communities" in Peru and China.27 Although it is difficult to establish with complete accuracy the full extent of this revival in collecting, a survey undertaken by the Rural Advancement Foundation in association with Jack Kloppenburg, Darrell Posey, GRAIN, and Accion Ecologica in 1995 found that over 200 different U.S.-based organizations or institutions had instituted new biological-collection programs in the decade from 1986 through 1995.[28] All of this raises a question: Why did this sudden, exponential increase in collecting occur at this time?

IMPETUS FOR THE REVIVAL: TECHNOLOGICAL CHANGE

The NCI's decision to reinvest in biological collecting, after abandoning its program only five years earlier, was critical in rekindling interest in and support for natural-products research in the United States. It is important, therefore, to understand what led to this sudden reversal in the NCI's strategy. The initial impetus for change was technological. Developments in the fields of molecular biology and genetic engineering in the late 1970s induced radical new approaches to drug design. Previously, researchers had relied on empirical "try and see" methods, in which thousands of different, randomly selected chemical compounds were tested for efficacy against a variety of disease states. This technique was laborious, haphazard, and economically inefficient—thousands of compounds extracted from plant or animal samples could be screened without producing any promising results. Two breakthroughs—the first in analysis of biomolecular structure and the second in high through-put screening were to revolutionize existing practices and provide the infrastructural basis for what became known as "intelligent" or "rational" drug design.

As molecular biology advanced in the 1980s, it became possible to explore the three-dimensional structure and the function of proteins such as hormones, enzymes, antibodies, and lymphokines that play key roles in bodily function and defense. Instead of having to guess at how drugs might impact on the body, researchers were able to generate much more refined knowledge of the structure of target molecules within the body. As the project to map the human genome advanced, thousands more molecular targets were identified.[29] Researchers were able to link this newfound knowledge of genetic function to information about the physiological effects of existing molecules in order to generate models of "ideal" drug molecules, allowing them to identify and target key proteins more effectively.[30] Two key information technologies—imaging technologies and computing technologies—were central to this project. Improvements in scanning electron microscopy and X-ray crystallography and the introduction of nuclear magnetic resonance imaging allowed scientists to visualize and map in three dimensions the arrangement of atoms within particular molecules, which in turn allowed them to model, with greater accuracy, the relationship or "fit" between receptor site and bio-active compound. The exponential increase in computational power simultaneously allowed researchers working on the design of new drugs to perform the literally millions of calculations necessary to predict an optimal relationship of "best fit" between compound and receptor. As ten Kate and Laird have noted, this computer-aided rational drug design drastically reduced the time required from finding a "lead" to identifying *and synthesizing* its chemical structure.[31]

The promise of these new technologies could not, however, have been realized without concomitant advances in the screening technologies that are used to establish the efficacy of compounds. From 1960 to 1980, the NCI had primarily screened biological material for anticancer activity using an in vivo system in which leukaemic mice were treated with various agents and then monitored for response. This process was extremely time consuming and costly. As Angerhofer and Pezzuto have noted, animal models were useful, but to achieve a successful isolation of a compound using them typically required an investment of a year's labor and more than 1,000 laboratory animals.[32] The emergence of AIDS in the early 1980s engendered a public-health emergency and increased pressure on the NIH to develop more efficient screening and testing protocols. The breakthrough came in 1986 when the NCI replaced in vivo screens with new in vitro screens involving the use of sixty human-cancer-cell lines.[33]

Although in vivo experiments continue to play an important role in the later stages of drug testing, the development of in vitro screening made it possible to conduct initial assessments much more quickly. The panel of sixty malignant human cell lines acts as a substitute for people suffering from different types of cancer: bowel, kidney, lung, and so on, while the extracted biochemical compound acts as proxy for the biological material from which it is extracted. Applying one to the other creates a very pronounced multiplier effect in terms of efficiency. Instead of having to wait for sixty human subjects to ingest substances and produce a reaction, the equivalent of those sixty subjects can be tested simultaneously, having been exposed via automated pipette delivery to a minute quantity of the compound of interest. By moving to in vitro screening, it became possible to reduce the amount of material required to conduct a test of efficacy by some 50–75 percent, which in turn increased the number of tests that could be undertaken using different protocols.[34] Eliminating many of the variables associated with in vivo testing also improved the accuracy of the testing procedure.

These factors combined to force a complete reassessment of the viability of natural-products research. As Gordon Cragg explained in an interview, these advances in screening revitalized the NCI's interest in biological collecting:

> In the early eighties the NCI dropped the natural products program because nothing new was coming out of it. But then in 1985, they started to introduce a new screening program . . . it was really with the introduction of that, that they decided to revive natural products because they thought, "Well, now we might be able to find something new."[35]

Advances in automation and roboticization, which facilitated the development of high-throughput screening techniques, also dramatically increased the *rate* at which compounds could be screened, enabling over 40,000 samples to be screened per year by 1990.[36] Improvements in cryogenic storage techniques were also having a dual effect. They were, first, enabling mass stocks of the human tumor-cell lines that were employed in the in vitro screening process to be stored and used as reservoirs for replacement of the cell panels when they became exhausted and, second, allowing collected

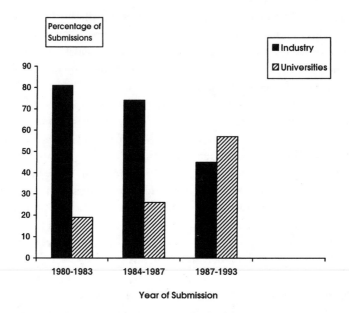

4.3 Submissions to the NCI's screening program, 1980–1993. Source: M. Grever, S. Schepartz, and B. Chabner, "The National Cancer Institute," 622.

samples of plant and animal material to be stored for future investigation and use. The effect that this could have on improving the efficiency of the collecting process was immediately apparent to those agencies the NCI had subcontracted to undertake the collections on their behalf. Doug Daly, the curator of Amazonian Botany at the New York Botanical Garden was responsible for overseeing the NCI contract. As he has noted, advances in cryogenic storage were critical in improving the economic viability of collecting, as they allowed the collected material to be stored for future rescreening, effectively extending its lifetime as an industrial commodity. As he explained:

> They suddenly had this ability to save the material that wasn't used immediately and to archive it at −20 degrees so it could be used again in the future without having to go to the expense and effort of collecting it all again. That was a major breakthrough. It made collecting a lot more cost effective because it was really difficult and expensive to have to do recollections.[37]

These technological developments also produced unexpected benefits for other collectors. The NCI had a long-standing policy of screening material for pharmaceutical companies, academic and scientific research institutes, and the like; however, use of this facility had historically been restricted to large pharmaceutical companies, as they were the only collectors who could supply the quantity of material necessary to conduct an accurate test of efficacy using old screening techniques.[38] The introduction of new screening technologies enabled researchers working with limited quantities of material to submit them for evaluation. This development radically changed the profile of suppliers to the testing program, with submissions from universities and other smaller private companies rising considerably (see figure 4.3). During this period (the mid- to late 1980s), scientific and academic institutions were also becoming increasingly conscious of the fact they could apply for patent protection for any novel pharmaceutical compounds extracted through this screening process. These factors created a further incentive for publicly funded institutions (such as universities, research institutes, and botanical gardens) either to begin collecting materials for commercial evaluation or, alternatively, to submit materials that they had already collected for testing.

THE BIODIVERSITY CONVENTION:
NEW PROTOCOLS AND NEW RATIONALES

Although these technological advances provided an initial impetus for the revival in collection, they cannot fully explain why this interest consolidated, and then escalated, over the following decade. The reasons for this are inevitably more complex and relate to changes that were occurring in the broader regulatory and economic environment. Despite the many laudatory articles and press releases that had accompanied Merck's very public announcement of its new bioprospecting agreement, news of the venture did not meet with universal acclaim.[39] As I suggested in chapter 3, the sudden rise in new collecting programs had begun to engender some serious concerns about the potentially "neocolonialist" aspects of collecting. Foremost among these were concerns that the collection and exploitation of genetic and biochemical materials would amount to a new form of "biopiracy." Unfavorable parallels were inevitably drawn between the NCI's new collection program and those enacted in the service of empire some 200 years earlier. The fact that the NCI's contracted collecting agencies—botanical gardens and museums of natural history—were, in effect, the modern counterparts of the old centers of calculation made such comparisons difficult to escape. This gave rise to precisely the kind of undesirable publicity that many of the large pharmaceutical companies were anxious to avoid.

These concerns were aired in the preparatory discussions for the negotiation of what would become the biodiversity convention and resulted in the introduction of new protocols that established the terms and conditions under which genetic resources would *ideally* be collected and used. These measures were designed to ensure that genetic and biochemical materials were exploited in "a just and equitable manner." Under the terms of the convention, genetic material became constructed as a resource in its own right. These resources would form part of the patrimony of each nation-state, and each would have the right to exploit them as it saw fit, pursuant to several new protocols. These protocols required those involved in the exploitation of such resources to utilize them in a sustainable fashion and to ensure that suppliers of genetic or biochemical resources (many in developing countries) receive appropriate compensation for their use, thereby, ideally, creating an incentive for their conservation.

The first Bush administration expressed concern that these provisions would harm U.S. competitiveness in emerging industries such as biotechnology and consequently elected not to ratify the convention. Although President Clinton went on to sign the convention in 1993, the Republican-dominated Senate declined to ratify it, citing dissatisfaction with the low level of intellectual property rights protection offered by the convention and its potentially damaging effect on domestic agricultural and pharmaceutical industries. Even if the United States had ratified the convention, it would have been under no legal obligation to implement the protocols, as they are not legally binding on the parties. In other words, the convention is a "soft-law" instrument that provides guidelines for action rather than enforceable legislation. In light of this, it is particularly surprising to discover that that the first attempts to translate these soft-law regulations on access and benefit sharing into formal binding contracts were, in fact, made by a U.S.-based collecting organization—the NCI. Stung by criticism that it had been their newly instituted global collection program that had initially sparked anxieties about "biocolonialism" and the monopolization of genetic and biochemical material, the NCI decided in 1988 to begin to devise a formal agreement governing the terms of access to and subsequent use of collected genetic and biochemical material. This agreement became known as a "Letter of Intent."

As the name suggests, this agreement was a statement of intent to commit to the principle of compensating source countries for the use of their genetic and biochemical materials. Under the terms of the agreement, compensation is disbursed in three stages. "Short-term" compensation comes in the form of an initial fee for collecting the material in the source country. This is usually paid directly to the intermediary organization undertaking the collection. "Medium-term" compensation takes the form of infrastructural support and training for technicians and scientists in source countries—usually those working for the intermediary collecting institution. "Long-term" compensation is secured (theoretically at least) through an agreement that obligates the NCI to require that any licensee of a drug produced from the material "seek, as it's first source of supply, natural products available from the source country . . . subject to the negotiation of a mutually agreed fair price" and/or provide a royalty to source countries on products produced from the collected materials or analogues thereof. In return, the NCI secures the exclusive right to patent "all inventions developed under this agreement" including all compositions of matter such as active agents derived from the material.

The LOI was later revised to become the "Letter of Collection" and later still the "Memorandum of Understanding."[40] Gordon Cragg was centrally involved in drafting this first agreement. There were a number of powerful incentives for him to lobby for its implementation. As he noted in an interview, there was entrenched resistance within much of the collecting community to the notion of introducing benefit-sharing agreements, and some contracted collectors had deliberately circumvented the requirement—a dubious practice that had tarnished the reputation of the program, creating a situation that he was anxious to remedy.

> Most of the grantees[41] are already very aware of these requirements, but as I say, there are those who I feel have abused the rights of the countries that they collect in—they might pay lip service to it, but in actual fact we know of instances where countries have been very unhappy, and that is what is promoting our interest in enforcing it [the LOI]. We know there has been a lot of unhappiness, and of course we know that it rebounds on us too because a lot of these folk, even though they are supposed to be independent, . . . get their funding from us, then of course when they abuse their privileges it comes back to us—it affects our actual programs—so there have been problems here, but hopefully we are going to get that under control. There is still an attitude particularly amongst the older generation [of] "you know, what are these countries sort of wanting all these rights for when really it's a global resource"—but that's changing slowly.[42]

Article 15, section 1 of the Convention on Biological Diversity affords countries the right to develop their own regulations governing access to genetic resources that reflect and accommodate regional specificities. Despite this, the LOI went on to become the model for most, if not indeed all, of the compensatory agreements that now govern the commercial collection and use of genetic and biochemical materials worldwide. The basic three-phase structure of the LOI devised by the NCI (with its reliance on short-, medium-, and long-term compensation in the form of an upfront collection fee, training and infrastructure, and a royalty on products) has quickly become the norm, proscribing a set of terms and conditions for the exploitation of genetic resources from which there has since been remarkably little deviation.

Despite the fact that none of the U.S.-based pharmaceutical companies or public institutions were required to introduce such compensatory agreements, quite a number were prepared to do so. Why was this? It could be argued that the desire to introduce such agreements reflected a wish to rectify past inequities, but it may, of course, also be explained by reference to their need to avoid adverse publicity. Organizations and companies were seeking a means to formalize and legitimize their collecting activities, and the convention, it could be argued, accommodated this requirement. It established a clear set of terms and conditions under which genetic resources might be "equitably" and "sustainably" exploited and provided a new "ecofriendly" rationale for undertaking biological collections.

One need only turn to a review of the press releases, promotional material, and articles that accompanied the launch of several new collecting programs in the late 1980s and early 1990s to see the important role that the convention and these new contractual agreements played in legitimating new bioprospecting activities. An analysis of twenty-seven pieces of such literature collated for this study reveals that all but one made frequent and explicit reference to the role that these new agreements would play in providing local communities with a source of ongoing revenue as well as a mechanism through which to conserve endangered environments.[43] For example, in describing the activities of his company for an article in *Worth* magazine in 1993, the vice president of Shaman Pharmaceuticals sought to draw attention away from the principal purpose of the company's bioprospecting activities—to secure raw materials for the industrial production of pharmaceuticals—by pointing out the apparently vital role that bioprospecting could play in preventing rainforest destruction:

> It's not just that Shaman develops drugs; the company also provides a market-driven rationale for preserving rain forests in South-east

Asia and the Amazon. By identifying harvestable products, Shaman gives local people a powerful reason to preserve the rain forests, now being cut at a rate of 129,600 acres per day.[44]

The same discourse was being employed to publicize the Merck–InBio bioprospecting agreement. Merck's promotional material focused attention on the role that the agreement could pay in providing ongoing economic returns and new incentives for the preservation of endangered environments in Costa Rica. This argument went on to occupy a central place in the numerous articles and media reports that accompanied the launch of this agreement. This was not an argument that Merck forced upon its collaborating partners. On the contrary, those working for the Costa Rican state government and in the InBio organization were among the most ardent supporters of the scheme. They, too, found these rationales useful in explaining to a broader public their decision to enter into a commercial bioprospecting agreement with a multinational pharmaceutical company. In defending the Merck–Inbio agreement at the time of its announcement, the directors of InBio drew heavily on these arguments. As one noted, "as these resources are becoming better appreciated by drug companies and other industries [they] are beginning to pay for the right to use natural products found in developing countries. Such moneys will be turned back into conservation of these biotic resources, thereby sustaining them in the long term."[45] Another argued, similarly, that "contracts between companies and collectors can help ensure that the exchange of biological materials generates both immediate and long-term benefits for source countries and communities."[46]

Indeed, as Bob Borris, one of Merck's senior research fellows, has confirmed in an interview, the impetus for introducing new regulations governing access and use came initially from source countries: "The decision to formalize what had previously been a fairly ad hoc system of collection was in fact made in direct response to developing countries' increased demand for permits and collecting contracts."[47] Some dissenting voices raised concerns about the terms and conditions of such agreements, but their criticisms had little impact on the prevailing orthodoxy of opinion.[48] Even in 1995, a full decade after the revival of collecting had begun, researchers at the University of Arizona who were collaborating on a new bioprospecting program were suggesting that the project "could result in the development of specialty cash crops for the country in which the compound originated. If a valuable chemical is discovered, for example, the source plant could be cultivated to provide local jobs or could be sustainably collected from the wild by people in nearby communities."[49]

It is difficult to underestimate how significant these new rationales were in combating entrenched notions about the inherently colonialist nature of collecting and in rehabilitating collecting as an activity that could play a key role in reducing global environmental destruction. Collecting and transporting valuable biological materials from lesser-developed countries to those more-developed but less-biodiverse centers of the world has always been a contentious activity. Charges that these collecting practices were inherently exploitative could now be countered by drawing critics' attention to the new protocols that had been introduced specifically to ensure that such practices were undertaken in a just, equitable, and sustainable manner. It was widely argued, as I have illustrated here, that these protocols would distinguish this new era of collecting from earlier colonialist ventures, despite any apparent similarity between them. By 1995, the "dual rationale" for undertaking bioprospecting—that it would create an incentive to conserve endangered environments by demonstrating the economic benefits that could accrue from their exploitation—had become canonical, taking on the force of fact on the basis of little more than hypnotic reiteration. The validity of this argument has never been tested. The question of whether these rationales retain any operative force will be analyzed in greater detail in the following chapters, where the efficacy of compensatory agreements will be closely scrutinized.

GATT TRIPS: NEW PROTECTIONS, NEW INCENTIVES

Although no references are to be found to it in the promotional material that accompanied the launch of these new collecting projects, another factor was providing an important catalyst for the revival in collection: the GATT TRIPs agreement. Establishing effective and internationally enforceable forms of intellectual property rights protection had become a priority within U.S. economic foreign policy as American corporations sought to protect the competitive advantages they had gained in high-technology industries such as computing, information processing, optical media (CDs, CD-ROMs, and DVDs), and biotechnology. The need to establish effective regimes of intellectual property rights protection for embodied inventions (those that find their expression in modifications made to life-forms) became particularly significant for the U.S. pharmaceutical industry in the 1980s as more and more drugs and therapies began to be produced biotechnologically. It was only eclipsed by the need to ensure that such regimes would be recognized globally, to curtail the unlicensed reproduction of patented materials or products in foreign, usually developing, countries.

Consequently, from early 1980s onward, the organization PhRMA (the Pharmaceutical Research and Manufacturers of America)[50] and its member companies began a sustained campaign, the aim of which was "to make the protection of American intellectual property rights a trade-policy priority."[51] As I noted earlier, attempts to extend U.S.-based intellectual property rights regimes to modified biological materials had already proved to be contentious, and attempts to adopt this regime as the basis of new world-wide trade-related system of intellectual property rights law met with considerable resistance. Countries such as India, Argentina, Brazil, and Turkey were particularly opposed to allowing patent protection for essential pharmaceuticals and for those that were developed using materials collected within their borders. Lobbying continued throughout the seven-year period of negotiation for the Uruguay Round of the GATT that extended from 1987 to 1994. It became increasingly evident during these years that developing countries would, in time, be required to recognize, adopt, and enforce a U.S.-based system of global intellectual property rights protection. Though under the terms of the TRIPS agreement, developing countries were granted a concessional ten-year period in which to institute these new regimes, the U.S. pharmaceutical industry, key U.S. government agencies, and the research-based industries of Europe and Japan continued to work closely together to seek accelerated implementation of TRIPS in key emerging markets.[52] By 1994, the United States had come to dominate production in the life-sciences industry globally, holding a greater percentage of patents for biotechnological health-care inventions both domestically and internationally than any other country or group of countries.

Another important indicator that reflects the changing context of biotechnological research and development in the late 1980s and early 1990s was the sudden increase in what are referred to as "Offices of Technology Transfer." Most large pharmaceutical companies established these offices (which are responsible for arranging patents and licensing and transfer of materials and technologies) over twenty years ago. What is interesting to note is the number of *publicly funded* institutions, such as universities and museums of natural history, that began to establish these facilities in the late 1980s. Interest in strengthening technology-transfer rights usually occurs when the context of research and development is undergoing a period of change, for example, when there is a need or opportunity to increase intellectual property rights protection, when cuts in government funding force new alliances with other institutions or corporations, or when individual researchers establish new contractual relationships with private firms.

Scientists working in publicly funded research institutions were becoming increasingly aware that the technologies and materials that they were developing (particularly those that were biotechnological in kind) might be eligible for patent protection, and they became more anxious to secure their rights over these inventions. Their desire to secure greater levels of protection increased as funding cuts led institutions into new partnerships with industrial corporations who were eager to secure access to the fruits of their knowledge and innovative work. Of the forty-five publicly funded research institutes that I surveyed for this study, some 30 percent had established their offices of technology transfer at some time from the mid-1970s onward but were uncertain of exactly when. A further 20 percent established them between 1985 and 1990, while another 30 percent established them in the following five year period.

Unsurprisingly, few of the researchers that I interviewed who were directly involved in the collection of materials were conscious of the effect that the ratification of GATT TRIPs would have on the pharmaceutical industry or on natural-products research. This is possibly because most work as scientists in intermediary agencies such as botanical gardens and museums of natural history and prefer to see their role as that of scientific investigator rather than as industrial subcontractor. When asked to comment on the impact that that the TRIPs agreement had on improving the viability of natural-products research, Michael Balick, the director of the NYBG's collecting program responded by remarking: "You know people want to come up and talk to me about the GATT—I don't know anything about the GATT. . . . It sounds convincing to me but you'd have to ask someone in business; I'm not in business."[53]

Of course, when researchers subcontract their services to industry that latter point becomes open to debate. While many collectors remain oblivious of the effect that the introduction of more robust intellectual property regimes have had on collecting activities, lawyers working within technology transfer offices are not. Conversely, they are acutely aware of the impact that the GATT TRIPs extension has had on increasing levels of patent protections for pharmaceuticals created from collected natural materials. As

Kate Duffy Mazan, a senior attorney in the NIH's Office of Technology Transfer explained, the GATT extension played an important if largely unrecognized role in increasing incentives to engage in bioprospecting:

> It sure makes it more commercial. The GATT is having the effect of creating harmonization in the patent policies of the member states, and therefore places that might not have allowed patents on pharmaceuticals now will. . . . Anytime you can get a patent right it gives you more exclusivity in the market—and that is an advantage—that is the whole point to the patent system—to make these things more marketable. . . . The effect of the GATT on bioprospecting has been primarily that it has created more countries with broader patent protection for things such as pharmaceuticals where natural products might play into that market. Basically it creates a much larger market for natural products.[54]

The combined effect of the introduction of the GATT TRIPs agreement and the biodiversity convention was to create a new regulatory and economic space—something as close to those much-vaunted "global spaces" as has yet been seen. This was a space within which genetic and biochemical resources became constructed and formally constituted as industrial commodities available to be engineered, patented, and exploited in "ethical," "just" and "sustainable" ways for what was ostensibly a greater global good—their future preservation. In short, it was an environment in which a new economy of trade in these bio-informational resources could both emerge and flourish. These factors alloyed with significant technological advances to create an enormous revival of interest in the collection of biological materials for applied pharmacological use.

But what effect did this have on the actual mechanics of collecting practice: on who was involved in collecting, on how they collected and perhaps most importantly of all, on what they collected? The technological, economic, and regulatory changes that I have described here were introduced progressively throughout the decade from 1985 to 1995. By examining the organization of the *last* major bioprospecting program to be initiated in this period, it is possible to discern the extent to which these changes had transformed collecting practice during these years. I am employing this program as an illustrative example not because it was the only such program carried out at this time, but rather because it exemplifies so well the way in which these changes altered the constitutional and operational organization of bioprospecting practice in these years.

THE PRACTICE AND PROCESS OF COLLECTING

New Collectors

The International Cooperative Biodiversity Groups program, which is one of the largest biological collection programs to be undertaken in this contemporary era, was developed in the early 1990s, when U.S. government support for and interest in bioprospecting were at an unprecedented high.[55] Building on optimism generated by the Merck–InBio agreement, a decision was made to invest government funds from three separate agencies, the NIH, the National Science Foundation and the U.S. Agency for International Development, in a new collaborative bioprospecting project. The stated aim of the project was to integrate "biodiversity conservation, sustainable economic activity and drug discovery."[56] The contention that bioprospecting could provide a means for the conservation and sustainable management of threatened ecosystems unsurprisingly occupied a central place in the literature that accompanied the launch of the ICBG program:

> Because biological resources which benefit local communities are most likely to be preserved, chemical prospecting or more specifically, the development of pharmaceuticals from natural products can be used to promote biological conservation by providing an economic return from sustainable use of the resources.[57]

Although the rationales were familiar, the organizational structure of this new collecting project was not. For the first time, a collecting program was to be based around the concept of collecting consortia. More sophisticated than the simple linear network of collecting agents that had dominated earlier NCI-funded collecting projects, these consortia were organized as linked series of hub and node networks (see figure 4.4). The changes that were occurring in collecting practices at this time are evidenced in the constitution and operation of these new consortia. The consortia comprised a range of actors—some had long histories of involvement in biological collecting activities; others did not. The consortia were significant as they played an important role in drawing, or "enrolling," into formal bioprospecting operations, those organizations—such as nongovernmental conservation or environmental-protection agencies—that had had no previous involvement with them. They also, necessarily, had the effect of creating formal linkages between these neophytes and other more "established" collectors (such as pharmaceutical corporations) and solidifying partnerships between academic and corporate collecting organizations. They acted to create an intensively interconnected, extensive, and interdependent network of agents, each of whom was dependent, for at least part of its livelihood, on income that might be derived from the collection, utilization, and marketing of genetic and biochemical resources. They generated important new avenues of exchange for these resources and, in so doing, played an important role in creating lucrative new markets for them. Funding was initially granted to five separate consortia, each of which had responsibility for collecting in a different geographic area.[58] Each consortium typically comprised a U.S.-based academic research institute, two or more publicly funded research institutes—such as natural history museums or botanical gardens located in both the United States and the source country—as well as a single nongovernmental agency (see figure 4.4). Although the initiative was funded by the U.S. government, each consortium was "encouraged" to include a corporate or industrial partner, as it was considered that this would "favor the efficient development of novel approaches to drug development."[59] Commercial development of the material would, it was suggested, create incentives to preserve endangered habitats and to identify sustainable methods for harvesting targeted organisms.[60] As I have illustrated here, government-funded agencies and private companies have both financed collecting programs before, and government agencies such as the NCI have also collected biological materials that they have later licensed out to pharmaceutical companies for further development. However, this was the first time that government-funded agencies and corporate interests had entered into a bioprospecting operation as joint partners.

By the time that this program was launched in 1993, the rationale that bioprospecting could play a crucial role in conserving endangered environments had become so widely accepted that it was used to legitimate the involvement, in commercial bioprospecting, of a group of organizations that might otherwise have been philosophically opposed to such activities. Conservation agencies, such as the World Wildlife Fund, The Nature Conservancy, the World Resources Institute, and Conservation International (all nonprofit, U.S.-based NGOs), that had no prior record of involvement in this industry and had in fact lobbied for noninvasive protection of endangered forests, were enticed to apply as partners in the new ICBG programs. Their perceived role, as one of the agencies suggested, was to facilitate the commercial usage of these resources by providing a "key link between forest people and their industrial partners."[61] Conservation International's director of conservation economics, Steve Rubin, justified their organizations' involvement in the scheme to their shareholders by arguing that the ICBG ini-

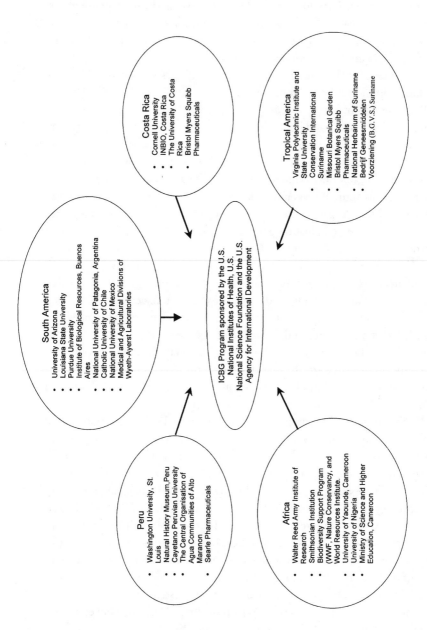

South America
- University of Arizona
- Louisiana State University
- Purdue University
- Institute of Biological Resources, Buenos Aires
- National University of Patagonia, Argentina
- Catholic University of Chile
- National University of Mexico
- Medical and Agricultural Divisions of Wyeth-Ayerst Laboratories

Costa Rica
- Cornell University
- INBIO, Costa Rica
- The University of Costa Rica
- Bristol Myers Squibb Pharmaceuticals

Tropical America
- Virginia Polytechnic Institute and State University
- Conservation International Suriname
- Missouri Botanical Garden
- Bristol Myers Squibb Pharmaceuticals
- National Herbarium of Suriname
- Bedrijf Geneesmiddelen Voorziening (B.G.V.S.) Suriname

Peru
- Washington University, St. Louis
- Natural History Museum, Peru
- Cayetano Peruvian University
- The Central Organisation of Agua Communities of Alto Maranon
- Searle Pharmaceuticals

Africa
- Walter Reed Army Institute of Research
- Smithsonian Institution
- Biodiversity Support Program (WWF, Nature Conservancy, and World Resources Institute.
- University of Yaounde, Cameroon
- University of Nigeria
- Ministry of Science and Higher Education, Cameroon

ICBG Program sponsored by the U.S. National Institutes of Health, U.S. National Science Foundation and the U.S. Agency for International Development

4.4 Schematic diagram of the ICBG network of collection, 1993–1998.

tiative would provide "ongoing opportunities to Surinam's economy and increased protection of its forests."[62]

Universities and government-funded research institutes also drew upon these rationales to explain and validate their growing involvement in commercial bioprospecting. While some of the larger institutional partners in the ICBG projects had prior experience collecting materials for commercial use, others, such as the Smithsonian, had none. These institutions had historically collected biological materials for purely scientific purposes, primarily taxonomy. The decision to enter into a formal subcontractual agreements with corporations to collect genetic and biochemical materials solely for use as an industrial resource marked a significant change of direction for many of them. Defending its involvement in the enterprise, the Smithsonian argued that such ventures could "encourage the preservation of tropical forests" and create a situation "in which everybody wins: the museum, the private sector, the host country and our society at large."[63]

Their stated concern for environmental protection aside, evidence from this research suggests that the primary motivation for these institutions to enter into these commercial collection projects is economic. U.S. federal funding for museums, botanical gardens, and biological research institutions began to decline in the late 1980s and early 1990s as the state began to cut back severely on support for research in the natural sciences. As the graph in figure 4.5 illustrates, U.S. federal funding for research in environmental biology dropped in 1985, flat-lined for the following five years, rose again briefly in the early 1990s, and dropped to the lowest level ever recorded in 1993. Although it was to recover somewhat in the following years, this period of continual underfunding was to have a particularly damaging effect on those institutions whose primary focus was the creation, maintenance, and study of biological collections.

In order to ascertain the effects of these funding cuts on institutions such as museums of natural history, botanical gardens, and university departments of biology and plant sciences, I surveyed eighty-four of them in 1996. Of these institutions, 42 percent had experienced a recent cut in government subsidy so substantial that they were forced to seek out alternative sources of funding for their research. One obvious possibility was to enter into contractual agreements to collect on behalf of corporate interests. This inevitably led to a dramatic increase in the number of institutions competing for a role in the ICBG program. Some thirty-four different consortia (each with several academic or institutional partners) were to apply for just five available ICBG grants.

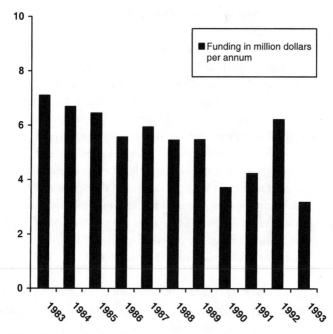

4.5 U.S. federal funding for research in environmental biology, 1983–1993. Source: National Science Foundation

As I noted earlier, approximately 30 percent of these publicly funded institutions opened technology transfer offices in this period from 1990 to 1995, an indication, perhaps, of their need to formalize their rights and obligations in relation to the external contracts that they were setting up with corporate partners especially in regard to patents on materials that they were developing in collaboration with industry. These new contracts and patents were to provide essential sources of funding to maintain existing research projects as rapidly declining government funding threatened to terminate them or suspend them indefinitely. Revenues from patented products, processes, and technologies have provided an extremely important and lucrative source of funding for institutions such as universities — Strathclyde University's Institute of Drug Research, for example, received over four million dollars of income in 1996 alone from royalties on patents.[64]

It would, of course, be naive to suggest that links have not traditionally existed between researchers working in public and private institutes, botanical gardens and universities, and pharmaceutical companies. It is evident that various informal collecting agreements have existed in the past between such agencies. However, new cooperative agreements such as the ICBGs have had the effect of formalizing and strengthening these existing but casual relationships while simultaneously encouraging the active participation in bioprospecting of groups that have had no previous connection with this activity. In short, they generated a greater degree of formal linkage between a greater variety of collecting agents. This created a network of entrepreneurs interested in collecting and trading samples of biological material that was more extensive and cohesive than any since colonial times. The role that these new linkages played in thickening up the trading networks that underpin the emerging economy in bio-information is discussed in greater detail in the following chapter. Before turning to that issue, however, I want to consider the effect that the introduction of these new, more intensive networks of collection had on the everyday practice of collecting.

New Collecting Practices

The first effect of the introduction of these new collecting networks was to extend the geographic scope of collecting operations. Individual entrepreneurs, small companies, academics, and scientists from the United Sates have undertaken pharmacological collecting in many countries of the world throughout the twentieth century; however, it was not until large institutionalized bioprospecting projects began after 1985 that a systematic, centrally organized, and globally extensive program of collection was formally established. As figure 4.1 illustrated, the NCI dramatically extended the scope of their collecting operations during the late 1980s, drawing samples of material from forty countries. The general revival of interest in biological collecting and improvements in collecting and testing methodologies also enticed a host of smaller pharmaceutical companies, private and publicly funded research institutes, and even individual entrepreneurs to begin undertaking small-scale collecting expeditions, primarily to countries in Latin America and Africa.

The second and arguably more important effect of these networks was to make collecting practices more intensive. While organizations such as the NCI had collected in a large number of countries, they had usually secured samples either by organizing brief expeditions to target regions or by establishing short-term collecting contracts with intermediary organizations in those regions. The new ICBG partnerships enabled bioprospectors to establish an *ongoing* presence in eight countries: Suriname, Chile, Mexico, Argentina, Costa Rica, Peru, Cameroon, and Nigeria. After embedding themselves in this way, they were then well positioned to establish closer relations with source-country organizations and institutions. This was undertaken, in part, with a view to devolving to them a much greater degree of responsibility for undertaking the collections.

There are two possible explanations for the implementation of this strategy—the first is that the bioprospectors sought to empower these native institutions by providing them with the infrastructure and training necessary to examine, classify, and collect biological materials independently of large U.S.-based institutions and corporations. An alternative reading is that it was simply more cost-effective and productive for the consortia to have collections undertaken by source-country institutions. Although indigenous communities and organizations within target countries have historically viewed foreign collectors with some degree of suspicion, the promise of "equal partnership" in new collecting ventures, ongoing economic returns, and opportunities for conservation encouraged them to become handmaidens to the bioprospecting industry, a role that they might otherwise have rejected.

These local intermediaries have clearly played a critical role in facilitating the collection process by improving field collection techniques, providing knowledge of local species and conditions, and assuaging local anxieties. The NCI has never sought to downplay their contribution, which Dr. Gordon Cragg described as "indispensable to the success of NCI's collection operations to date."[65] A number of organizations in host countries, such as InBio in Costa Rica, actively sought increased involvement in bioprospecting projects supposing, that this would provide an important opportunity to secure the advanced technologies and training necessary to improve domestic research capacity.

It is interesting to reflect on the motivations for transferring technologies and training to these developing countries, and I shall do so further in later chapters. For now, it is sufficient to note that much of the funding that was originally offered to InBio by Merck was allocated to the training of taxonomists, of which there were reportedly only ten in Costa Rica in the late 1980s.[66] InBio set about enlisting local people, including "farmers, ranchers, high school students, housewives and national park guards," to be trained as what are termed "para-taxonomists."[67] Although many had little or no formal education, these trainees received a general introduction to biology, ecology, and field techniques, providing them with the skills necessary to collect specimens, mount them, and undertake a preliminary sorting of them into species groups. Parataxonomists also performed simple field and laboratory experiments and observations and undertook basic biodiversity survey work. The parataxonomists that InBio trained were employed by them to simultaneously inventory *and collect* specimens of biodiversity. One set of samples was collated to form part of a wider "Biodiversity Inventory"—a reference collection of the countries' biodiversity—while another set was deposited in InBio's repository for investigation and screening by Merck. It was reported in 1990 that the parataxonomists from the first training course were successfully collecting in excess of 100,000 insect specimens per month.[68]

Other indigenous groups and communities have been more reluctant to participate in collecting programs or to provide information to collectors about local conditions, species of plants or animals, or the medicinal uses to which they might be put. Although random sampling and ecorationalist[69] approaches were popular collection methods in the 1960s and 1970s, they fell out of favor in the 1980s as a consequence of their high failure rate.[70] It was argued at the time that the success rate for the identification of useful bioactive agents could be dramatically improved by employing a process known as "cultural pre-screening." This process involves utilizing generations of indigenous knowledge about the medicinal uses of plants and local people's knowledge of the plants in their environment to more effectively identify those with proven efficacy.[71] The presumption is that by acquiring indigenous knowledge about the uses of biological materials, collectors can target their searches for pharmacologically active materials, allowing them to produce a greater number of compounds from these screens than would have been produced from samples collected at random.

Securing the cooperation of indigenous groups in this endeavor has proven to be a complex task. Despite various inducements, many remained concerned that their resources and knowledge would be expropriated without their consent or appropriate compensation. It became essential for companies and institutions to identify new methods of involving disaffected communities, and they consequently began to turn their attention to the work of ethnobotanists. Ethnobotany, which was defined by its founder as the "the study of plants used by primitive and aboriginal people" was established as a formal academic discipline by American botanist John. W. Harshberger in 1895.[72] Interest in the study of ethnobiology had increased markedly in the mid 1890s, having been stimulated by exhibitions on the subject that were shown by anthropologists and archaeologists at the 1893 World's Fair. The first doctoral dissertation in ethnobotany was awarded in 1900 by the University of Chicago, and, by 1916, the discipline had expanded to include studies of applied uses of plants and other biological materials, as well as analyses of the way these materials and uses were perceived and understood within different cultures. The data that was generated through these studies formed the basis of new subdisciplines such as ethnopharmacology.

In recent years there has been a marked resurgence of interest in the applied uses of ethnobotanical knowledge. In the field of bioprospecting, this interest was initially piqued by a report given at a symposium on economic botany held in Bangkok in 1990. The report, which was authored by Michael Balick, now director of the Institute of Economic Botany at the NYBG, indicated that an appreciable increase in the number of biologically active compounds could be obtained when employing an ethnobotanically directed, rather than random, approach to plant collection.[73] This finding, which was widely reported, convinced corporations and institutions that searches might be more efficiently organized when directed by an ethnobotanist, and ethnobotanists consequently became more directly involved in bioprospecting operations. The introduction of a commercial imperative would, however, inevitably alter the nature of their involvement with indigenous communities.

The ethnobotanists' role was no longer simply to document specific medicinal uses that were made of collected materials but also to actively persuade native peoples that they would benefit from involvement in these bioprospecting programs. Employed increasingly by large corporations, companies, and institutions for this explicit purpose, ethnobotanists themselves began to redefine their role, with one prominent authority suggesting in 1990 that they were professionals with both botanical and anthropological training, "responsible for identifying the healers in an indigenous culture and securing their participation in the [bio-prospecting] research program."[74] They met with considerable success in these endeavors in the early 1990s, recruiting an increasing number of indigenous and tribal groups to the global bioprospecting initiative. Several of the largest bioprospecting operations that were undertaken in the late 1980s and early 1990s, including Shaman's operations in Peru and Ecuador, the NYBG's program in Belize, and two of the five ICBG projects, have been based on ethnobotanical research. None would have succeeded without the close involvement and cooperation of local indigenous groups. The impact that this intensive engagement has had on particular communities is, of course, varied and complex, as Hayden's and Green's more detailed ethnographies reveal.[75]

The scope and intensity of this recent wave of bioprospecting remain unprecedented in the history of the U.S. pharmaceutical industry. More collecting programs have been introduced than ever before, a greater range of actors are involved, collections are being undertaken in more and more countries. Economic and regulatory changes that occurred during the late 1980s and early 1990s combined to construct genetic and biochemical materials as alienable commodities and create conditions that favored the establishment of a new market or resource economy in biological derivatives. They created new rationales and incentives for private and publicly funded organizations and institutions to become actively involved in biological collection, creating a global network of collection agents that act as conduits through which collected samples are conveyed from the periphery to what are the contemporary equivalent of Latour's "centers of calculation." While these factors transformed the way in which collecting programs were organized and conducted, other technological changes were simultaneously revolutionizing the actual mechanics of the collecting process, altering what was collected and how.

New Collections

During the early 1980s, the focus of biological research began to turn inward as advances in molecular biology and new genetic engineering technologies led researchers to undertake ever more detailed examinations of the component parts of organisms. Researchers interested in investigating the medicinal uses of plants and animals have long collected specimens—ranging from whole organisms (such as plants or insects), to parts of organisms (collections of leaves or roots), to substances that had been extracted from whole organisms (such as venom or sap)—with the aim of testing them for pharmacological activity. With the emergence of new approaches to this study and the introduction of associated technologies in the 1980s, these collection practices became much more sophisticated. Biological materials could be rendered in new forms and this, coupled with revolutionary new methods of storing, screening, and utilizing these new proxies, began to progressively transform the social and spatial dynamics of trade in these materials—providing collectors with a more efficient means of collecting, concentrating and controlling, and recirculating to their advantage just those key or essential elements of biological materials that they most valued. The significance of these developments is most clearly evidenced by charting the changes that have occurred in the way in which biological materials are collected, transported, and stored.

In instances where the material to be collected is a relatively small organism, such as a microbe, insect, fungus, or small marine organism, it often remains as efficient to collect the whole organism as it is to collect a sample of that organism. The Costa Rica ICBG and the Merck–InBio agreement both involve the collection of whole insects, while the NCI and Bristol Myers Squibb have collected fungi and microorganisms found in soils. Smith Kline Beecham and the NCI have also been involved in the collection of whole marine organisms, such as kelp, sponges, invertebrates, crustaceans, and corals, through their partnerships with the Smithsonian Oceanographic program and the International Coral Reef Foundation, respectively. Although the collection of whole organisms is not as commonplace as it once was, it does still occur. The collection of whole organisms is, however, a practice that attracts some trenchant criticism. Serious concerns have been raised about the number of organisms that are being harvested and the possible impact that this might have on breeding populations. *The New Scientist*, for example, reported in 1995 that a bioprospecting group from the United States had collected *450 kilograms* of the acorn worm *Cephalodiscus gilchristi* in order to isolate *1 milligram* of an anticancer compound.[76] Such activities also raise important questions about how the extracted compound will be used. As such a rate of harvest is clearly unsustainable in the long term, it would appear that any future industrial use of the compound would require the development of a synthetic analogue—an issue I discuss in greater detail in the following chapter.

The collection of larger whole organisms (higher plants and animals) has always been an unwieldy business, and field collectors were encouraged to obtain an appropriately sized sample on which to conduct tests. As I noted earlier, advances in screening technologies introduced at the NCI in the late 1980s were to dramatically reduce the amount of material required to undertake such tests. Perhaps surprisingly, this did not, at first, act to greatly reduce the size of collected samples. Although only milligrams of material were required, field collectors were still encouraged to obtain between 0.3 and 1 kg of plant or animal material. The material that was not used for testing was archived for future use in a new repository in Frederick, Maryland. During the late 1980s and early 1990s, the NCI required its contractors to collect 1,200 such samples of plant material per year.[77] While it might be assumed that the collection of a "sample" is more sustainable and less destructive than the collection of a whole organism, this is not always the case, particularly where the "sample" is a body part essential for continued function. *The New Scientist* reported another case of marine bioprospecting in 1995 in which 847 kg of moray eel *liver* was used to isolate 0.35 mg of ciguatoxin for chemical study. As one marine biologist remarked at the time, "just imagine how many moray eels were collected."[78] In such instances, the advantages of collecting a sample rather than whole organisms are purely economic—livers are easier to transport than are whole moray eels.

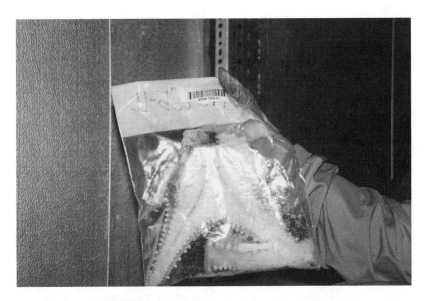

4.6   Processes of rationalization: The creation of highly characterized bar-coded specimens.

These collected samples of material then undergo an important process of in situ rationalization. It is at this point, for example, that the collected material becomes formally constituted as a specimen. This task is usually undertaken by a collaborating institution in the source country, for example, a state-funded herbarium or university research institute. In-country scientists and parataxonomists provide the initial technical labor necessary to transform "undomesticated" whole plants and animals into useful proxies—characterized samples. They play a crucial role in ordering the material in accordance with a set of received, universal norms and standards, stabilizing and mobilizing both the organism and information about the organism in such a way that both are rendered intelligible to those who will experience them only "at a distance." The material gains a further value through this process of rationalization, being transformed from an uncharacterized plant or insect languishing in situ to a taxonomically classified specimen that may be relocated through an attached documentary record of the exact temporal and geographic context (point of longitude and latitude, habitat, season, date, and so on) in which it was found. This information proves indispensable for explicating the relationship between the collected material and its environment once the material becomes decontextualized. This information, along with a record of the collector and a unique collection number, are expressed on a bar-coded label assigned to each specimen (see figure 4.6). These specimens are then air-freighted to repositories in the United States.

As testing and screening protocols improved, it became evident that what collectors most required was not whole organisms, or even samples of them, but rather the biochemical compounds and molecular substances latent within them. These were the "essential" or "key" elements of biological material that scientists had to have in order to conduct tests of efficacy—the rest of the organism could be divested if necessary. Extracted biochemical compounds became a proxy for whole organisms and samples, and, like other proxies, they proved to be much more lightweight and mobile than the bulk materials from which they were drawn. Large corporations and institutions had processed collected whole organisms and samples in their laboratories in the United States, but from the late 1980s onwards they began to invest considerable sums of money in developing extraction facilities in source countries. Enabling the collected materials to be translated into these new artifactual forms in situ greatly improved the ease with which valuable genetic and biochemical resources could be transported, allowing far greater quantities of such materials to be conveyed in any one shipment.

While improvements in screening technologies were reducing the amount of material required to test for efficacy, two further biotechnological advances in the late 1980s and early 1990s combined to extend the viability of collected materials. It has long been possible to extract compounds from whole organisms or samples using various chemical or aqueous solutions; however, these extracts were often unstable and did not store well. In recognition of this, the NCI began in the late 1980s to develop new processing protocols that were designed to "maximize the probability of extracting biologically active molecules from specimens and of *preserving* that activity through the processing and storage of the extracts."[79] These new techniques were successfully introduced in tests on plant and marine samples in 1988.

Linked to this were advances that were occurring in the application of cryogenic storage techniques. With the introduction of more sophisticated molecular approaches to drug development, tissue preservation protocols also began to undergo a necessary transformation. Many of the samples of material that had been collected for investigation on earlier expeditions had been preserved by drying or through immersion in formalin or ethanol, procedures that severely degraded the molecular structure of the organism, prohibiting any study of sensitive biochemical interactions and functions. It quickly became evident that new genomic technologies would require high-

quality tissue resources. Voucher specimens that had historically been collected, as one researcher put it, "at the end of a shotgun" were of comparatively little value.[80] There came a realization that to collect a specimen that
had only been analyzed on a morphological and not molecular level was the
equivalent of "keeping the wrapper and throwing away the prize."[81] In order
to conduct the kind of detailed molecular and cellular investigations upon
which research in the life sciences would come to depend, it became necessary to cryopreserve specimens and tissues in the field as quickly as possible after harvesting them. Portable, hand-held cryogenic storage flasks, or
"dry shippers" as they are known, which were first used industrially in the
United States to transport bull semen in the 1950s, were introduced into
natural-products research in the early 1990 and are now routinely employed
by field collectors in any circumstance where maintaining enzymic activity
or molecular integrity is paramount.[82]

    These cryopreserved materials, and other collected specimens of tissue,
are then shipped back to the United States, where they are transferred to
much larger cryogenic freezers within specialized repositories and archived
for future use (see figure 4.7). Several such repositories exist in the United
States. Some, such as the American Type Culture Collection have been in
existence since the 1930s. The ATCC is a nonprofit bioresource center
based in Washington, D.C. Part of its stated role is to "acquire, authenticate,
preserve, develop, and distribute biological materials and by-products, to private industry, government, and academic organizations around the world."
The ATCC holds vast collections of microorganisms, cell lines, and recombinant DNA materials, viruses, phages, bacteria, plant-tissue samples, and
many other biological materials. The plant collection alone contains some
27,000 strains of fungi and yeasts, 75 plant-cell cultures, and over 400 strains
of patented seeds. All of these collected materials are housed in a state-of-
the-art repository that comprises over 8,200 square feet of storage space. The
ATCC has been at the forefront of advancements in cryopreservation storage techniques for many years, and their repository facility contains fifty-five
ultralow mechanical freezers and some sixty-five vapor-phase liquid nitrogen
freezers. Although they work at the cutting edge of this field of endeavor, the
researchers at the ATCC have themselves been amazed at the capacity of
new cryogenic storage techniques to extend the temporal viability of an ever
increasing range of organisms. Bill Niermann, a former director of research
at the ATCC, and now vice president for research at the Institute for Genomic Research, described for me the extraordinary advances that were being made in cryopreservation in the mid 1990s:

4.7A Cryogenically stored marine tissue and plant samples.

We are preserving things by different methods. Probably our best method is in a vapor base of liquid nitrogen and basically nothing happens at that low a temperature and I don't know that anyone has yet explored the limits of viability for that. I suppose the limits of viability are simply how much genetic damage the material will experience from cosmic radiation, as there is no chemical reaction going on. We also freeze-dry many things for preservation. We freeze-dry them and then seal them under vacuum, and, for that, we've got documentation of greater than thirty years viability and liquid nitrogen is undoubtedly better than that so they could be viable for a very long time after that. There are even some protozoans

4.7B   Cryogenically stored marine tissue and plant samples.

that people are working on now, developing protocols for drying *protozoans even*—protozoans are single cellular organisms![83]

These advances had important implications for natural-products research, allowing collected materials to be stored almost indefinitely for future reutilization. As the director of the NYBG's Institute of Economic Botany explained to me, while some biochemical compounds break down or become volatile over time, others remain remarkably stable. As he noted, Richard Shultes's work on one-hundred-year-old *Banisteriopsus* samples revealed that the alkaloid content of this hallucinogenic plant had remained virtually unchanged over time—leading him to conclude that "they [biochemical compounds] don't disappear, that for the most part they can last a *very* long time."[84] Cryogenically storing these extracted compounds enables them to be remain tenable as industrial commodities for even long periods of time—possibly several hundred years. It is difficult to predict the possible value of these archived materials—their potential remains latent. New technologies are continually altering the uses to which collected material may be put, and it is impossible to envision now what those uses may be. As one industry insider reminded me, "screening methodologies are improving incredibly quickly—what may have absolutely no potential today may have extraordinary potential for tomorrow."[85]

New Collection Spaces

The generation of these new proxies has bought about a demand for new collection spaces. The task of transferring exotic collected specimens over great distances with the aim of reproducing them in new, and often environmentally unsuitable locations is a relatively complex one, requiring specialized technologies and expertise. During the sixteenth, seventeenth, and eighteenth centuries, a number of facilities—zoos, botanical gardens, and glasshouses—were purposely built to aid this project. Each provided a controlled "life-world" within which whole collected specimens could be maintained and reproduced. These collection spaces have remained relatively unchanged for generations. However, with the introduction of these new technologies it has become possible to store reproducible genetic and biochemical materials in new forms—as proxies such as cryogenically stored tissue samples, cell lines, blood samples, and biochemical extracts. The new repositories that have been built to house these proxies are similar to the old in that they, too, provide a space within which collected materials can be sequestered until they can be successfully redeployed. In other respects, the form and function of these new repositories reflect the differences that exist between the form and function of collections of biological proxies and those collections of whole living organisms from which they were derived.

One of the largest repositories of biological proxies to have been generated for use in the development of pharmaceuticals is the National Cancer Institute's Natural Products Repository. As I noted at the beginning of this chapter, the NCI had destroyed most of its earlier collections of samples in the early 1980s, as it believed that the potential of these materials had been exhausted. As new screening technologies broadened the possible range of tests to which collected materials could be subjected, the NCI came to rue the fact that it had not held them longer. As the biotechnological revolution really began to accelerate in the late 1980s and early 1990s, it became evi-

4.8   The NCI Natural Products Repository, Frederick, Maryland.

dent that the range of uses to which collected materials could be put might exceed any current imaginings. The NCI decided not to risk a repetition of its last, expensive failure of foresight. Conscious of the impact that improvements in cryogenic storage could have on extending the viability of collected materials, the institute decided to build a new and sophisticated facility in which to store samples of material collected under their new bioprospecting programs.

Although it is one of the largest such facilities in the United States, remarkably little is known, publicly, about the NCI's Natural Products Repository. Built in 1986, the repository is located in a nondescript brown warehouse in Frederick, in the heart of rural Maryland (see figure 4.8). Despite its unprepossessing exterior, the repository houses one of the most extensive collections of genetic and biochemical material in the world. Unlike the zoos, botanical gardens, and glasshouses of the seventeenth and eighteenth centuries, which had to have adequate room to accommodate whole living specimens, this facility is built to house proxies—and they have no such requirements. These new repositories are also devoid of the trappings of display; this material is not for public consumption. Rather, these are unashamedly industrial spaces, filled almost entirely with the twenty-eight

4.9　The cryogenic ark: Twenty-eight walk-in cryogenic storage freezers at the NCI's repository.

double-decker, walk-in cryogenic storage freezers that house the collected materials and the fail-safe machinery and systems required to maintain them (see figure 4.9).

In some regards, the collections resemble the floating herbaria that were found aboard the earliest ships of discovery, with samples neatly ordered and arranged. However, these collections of material are distinguished from their historical counterparts in one vitally important regard. This contemporary collection of material can be used not only for purposes of taxonomic identification and comparison, but also as a source of readily replicable and manipulable genetic and biochemical material. These materials remain vital and potent, in other words, unlike the dried and pickled specimens of old.

Progressive processes of rationalization have allowed collected whole organisms to be rendered as cryogenically stored tissue samples, biochemical extracts, cell lines, and the like—all of which retain full enzymic and cellular function. This process of "distillation" allows collectors such as the NCI to maintain viable, reproducible genetic and biochemical materials in a much more condensed form and, thus, in a much more condensed space. The NCI collection housed at Frederick consists of tissue samples from

4.10A "Collecting the world": Some of the thousands of cryogenically stored plant biochemical extracts and tissue samples archived at the NCI's repository for future use.

some *50,000 organisms*, which the director of natural products research estimates may represent between 15,000 and 20,000 different species. Over 114,000 different biochemical extracts have been refined from this material, and all of these proxies — both the cryogenically stored tissue samples and biochemical extracts — are archived in the repository for long-term future use. It is difficult to imagine how it would have been logistically possible for any collector to amass (and transport, store, and maintain) an equivalent number and type of whole living organisms. Changing the way in which these genetic and biochemical materials are rendered and stored has clearly facilitated the collectors' ability to concentrate these increasingly valuable commodities for use over time (see figure 4.10).

Of course, the NCI's repository is not the only such establishment in the United States. An increasing number of others have been developed over the past two decades to house collections of genetic and biochemical materials, seeds, and whole specimens for agricultural, horticultural, taxonomic, and conservation purposes. In the pharmaceutical industry, however, it has only been the largest corporations that have had the resources necessary to build and maintain repositories of biological materials. The largest privately funded collection of biological material used exclusively as a resource for

4.10B   "Collecting the world": Some of the thousands of cryogenically stored plant biochemical extracts and tissue samples archived at the NCI's repository for future use.

pharmaceutical development is owned by the pharmaceutical corporation Merck. While other pharmaceutical companies, such as Smith Kline Beecham, Bristol Myers Squibb, and Pfizer, have amassed collections of material, these are usually stored for them by the research institutes that undertook the collections on their behalf. Merck is unusual in that it prefers to maintain its own collection of cryogenically stored samples and extracts at its headquarters in Rahway, New Jersey. Although Merck would not reveal the exact size of their collection, one of their senior researchers did confirm that it rivals the NCI's in size, consisting of "multiple thousands of samples of different organisms."[86] In addition to this, they have an herbarium collection that is "stored at a botanical garden . . . and a library of extracts of plants" that are, in his words, "living at −20° C, a few doors up."[87]

Universities, museums of natural history, and botanical gardens have all been contracted to undertake field collections on behalf of large pharmaceutical companies in recent years, and many hold small to medium-sized collections of specimens of plants or animals or even, in some cases, libraries of biochemical extracts collected on their behalf. For example, the New York Botanical Garden and Cornell University's entomology department both hold collections of material collected on behalf of the pharmaceutical company Merck. Academic and scientific organizations such as these have a long history of collecting specimens for research purposes, and many already have large collections of living or fixed materials. Materials collected on commercial bioprospecting expeditions should ideally be maintained separately from existing collections, as the material has been collected for different purposes and, therefore, usually under different terms and conditions. Whether all institutions maintain this distinction is not known. The fate of existent ex situ collections of material is of particular interest in light of the renewed demand for samples of biological material and will be investigated further in the following chapters.

One last point needs to be raised in relationship to these new collection spaces, and it pertains to the nature and life span of the collection spaces themselves. In recent years, cryogenic storage facilities have become much more transportable and affordable than they once were. Numerous small pharmaceutical companies such as Shaman, small biotechnology-supply companies such as Pharmacognetics , individual academic and scientific researchers, and entrepreneurs have consequently acquired the technology necessary to create their own small collections of cryogenically stored materials. Many now store collections of tissue and libraries of extracts in relatively small refrigeration units that can be fitted into any commercial or even domestic space. At a minimum, this is the only specialist environment required to maintain the bio-informational resources embodied in the extracted genetic and biochemical materials. Collection spaces are no longer necessarily either dedicated or enduring in the way that zoos, botanical gardens, or even glasshouses were. The task of tracing the fate of collections is complicated by the fact that collection spaces themselves may prove to be as mobile as the materials they now house.

Accelerating the Social and Spatial Dynamics of Collecting

Although it may appear unchanged, collecting practice has in fact undergone a subtle process of transformation over the last two decades. Evidence presented here suggests that these changes are both linked to and driven by technological advances and by changes in the global economic and regulatory environment. These changes have combined to spark a resurgence of interest in the project of collecting biological materials for commercial or industrial uses. Biotechnological advances have played a particularly important role. They have had the effect of enabling genetic and biochemical material to be rendered in a variety of new, more transmissible, artifactual forms, allowing them to be transported and used independently of the organisms in which they originally existed. Although it is still possible, and in some cases desirable, to collect whole living organisms for investigation or reproduction, it is no longer *necessary* to do so. Most collectors are now principally interested in accessing the genetic and biochemical materials embodied in whole organisms—which, in the case of the pharmaceutical industry, may be used as lead compounds for drug development or as the basis of new gene therapies. Collectors need no longer acquire breeding populations, or even cumbersome whole plants or animals, but rather can be satisfied with a variety of proxies that will bear the genetic and biochemical material or information required.

The first effect of these changes on the social and spatial dynamics of collecting is clearly apparent: as a consequence of these changes, the genetic and biochemical resources embodied within biological material have become infinitely more amenable to collection and to concentration. A greater number and more types of species can now be collected and transported from one location to another and concentrated within a surprisingly small space and in a very accessible form. At the same time, concomitant advances in screening technologies and in cryogenic storage techniques have had the combined effect of reducing the amount of material that is required to constitute a viable industrial resource, while allowing that material to be maintained for ever longer periods of time.

New global protocols introduced in two forums, the GATT and the biodiversity convention have also had the effect of creating new incentives to collect: the biodiversity convention by constructing genetic and biochemical materials as alienable resources there to be "sustainably" developed; and GATT TRIPs by providing, first, an unprecedented degree of patent protection for inventions based on modified genetic or biochemical material and, second, by creating a new regulatory space within which these rights are viewed as normative and requisite of universal recognition. That these developments have had the effect of creating new incentives to collect is clearly evidenced by the increase in the number and different types of organizations that have become involved in bioprospecting over the last decade. The creation of new strategic alliances and interlinkages between a host of participating organizations has had the effect of creating extensive new networks for the collection, use, and exchange of bio-informational resources.

At one level, collecting these materials and concentrating them in particular localities seems an unproblematic practice. These materials (samples of plants, animals, even humans) remain present in the public domain and available for general use. Or do they? The materials that are concentrated here, are, I would argue, valuable precisely because they are comparatively exotic and rare. The plants and animals from which they are derived may not be rare, but the bio-informational proxies (those particular artifactual renderings of extracted genetic and bio-chemical materials that have been generated out of them) are. Part of their increased value must derive from the fact that these proxies can be utilized and transacted in novel ways—and from the fact that comparatively few groups or institutions within society have the power to utilize them in those ways. In the following chapters, I reveal how particular groups are capitalizing on the social and spatial dynamics that characterize the operation of this brave new bio-informational economy, and I reflect on the implications that this has for those who first provided the biological materials from which these valuable informational resources are drawn.

5. THE FATE OF THE COLLECTIONS

The resurgence of interest in collecting biological materials, which began in the 1980s and culminated with the introduction of so many new bioprospecting programs, set in motion a new "cycle of accumulation." Biological materials are, once again, being collected in many parts of the world and concentrated in dedicated repositories located now, as they have been historically, in the principal scientific research establishments of the developed world. Many of the largest of these repositories are now to be found not in Europe but in the United States, in what might be termed "New World centers of calculation." Of course, institutions such of these are not characterized as centers of calculation simply because they are centers of accumulation. If this were the case, the institutional laboratories and repositories that act as storehouses for collections of biological materials in developing-world centers would warrant inclusion as such, when in fact very few do. Simply holding a collection of biological material affords the collector no particular scientific or commercial advantage unless that material can be rendered and acted upon in novel ways. What makes corporations like Merck or institutions such as the NCI deserving of the title "new centers of calculation" is the fact that they have the technological capacity to explore and exploit the collected materials in ways that other collectors cannot.

In this chapter I set out to examine how technological developments have changed the way biological materials are used as industrial commodities within the U.S. pharmaceutical industry. Although a great deal of publicity attended the announcement and subsequent development of bioprospecting programs, almost nothing is known about the fate of the collections of material generated by these programs. Where were these materials taken, and how have they since been used? Who has benefited and who is likely to benefit from the exploitation of these materials and how? In an earlier chapter, I argued that biological materials are coming to share particular attributes with other informational resources: the abilities to be more readily conveyed or transmitted, unproblematically replicated across space and time, and easily modified to produce new versions.

Although few attempts have yet been made to establish the fate of these collections, I intend to do so here, in part because I believe that the absolute significance of the changes that have occurred in collecting practice over the past two decades are only revealed when more is known about how these collections of material have been used, traded, and exchanged. By drawing on findings from my investigation into the fate of these collections, I intend to argue and illustrate that there is a grave risk that the existing global trade in biological materials for life-sciences research (which provides economic returns for many communities) is being progressively undermined by the emergence of an untraceable, and hence unrecompensable, trade in "bio-information." In making this argument, I pay particular attention to revealing how new bio-informational proxies are used within the New World centers of calculation and circulated as commodities around the networks of exchange that underpin this new resource economy. By linking these findings back to the typology of the social and spatial dynamics of collection that I set out in chapter 2, I hope to illustrate how the creation of these bio-informational proxies has served to enhance greatly the ease with which genetic and biochemical resources can be, first, concentrated and controlled and, second, recirculated to advantage through processes of trade and exchange.

FROM REPRODUCTION TO REPLICATION

The process of producing a drug from a biological material may be time consuming and risky, but the mechanics of development are relatively straightforward. Once a biochemical compound has been extracted from a collected sample of material using a process of fractionation, it is then screened for activity against a panel of diseased cells. Should an agent exhibit sufficient activity, it is selected for further preclinical and clinical development and may eventually be produced commercially. A sample of collected material usually provides enough pure extract to undertake these initial screening processes. Further quantities are then required for clinical trials, and an ongoing supply of material must be secured once the drug moves into commercial production. The contention that source countries will benefit from this need for ongoing supplies of raw material has formed a central part of the justification for engaging in bioprospecting. It has been widely argued, as I illustrated in the previous chapter, that companies would return to source countries to secure large quantities of these materials either through in situ collecting or the development of extractive reserves or cultivation plots. These processes, it is argued, will create local employment opportunities and economic returns and thus an incentive to preserve endangered environments. Much has been made in the bioprospecting literature of the gains these activities could bring to local communities.

There is no question that further quantities of material will have to be procured if a drug goes into commercial production. The question that is rarely addressed however, is this: Further quantities of *what* material? The answer is quantities of the genetic sequences or biochemical structures on which the pharmaceutical is based—which is *not* to say quantities of the biological materials from which they were drawn. As we have seen, these genetic and biochemical substances enjoy a newfound sovereignty and mobility; they can now be extracted from organisms and stored and used independently of them. It is now, therefore, entirely possible to generate more of these substances without ever having to go back and collect or cultivate more of the organisms from which they were initially derived. Collecting organizations such as the NCI have continued to argue that the process of drug development "usually requires the procurement of *large quantities of raw material* for bulk production of the agent," for which they will "seek as a first source of supply, natural products from the original source country."[1] Evidence from this research suggests, however, that companies are increasingly finding it more expedient and more economically efficient to simply replicate the genetic or biochemical materials that they require in the laboratory from existing samples of collected biological materials.

The transition from biological reproduction to laboratory-based replication of genetic and biochemical materials has been a gradual and subtle one. "Reproduction" and "replication" are often used interchangeably in everyday conversation, suggesting that there is little functional difference between the two terms or, indeed, the two acts. While this may be true in some spheres, it is not so in biology. Reproduction in biology involves the generation of offspring through sexual or asexual means—with all the temporal and physiological constraints that attend such processes (need for species compatibility, fixed times scales of gestation, and so on) Replication, on the other hand, involves making an exact copy or duplicate of an existing biological "work" (a cell or gene) through more technologically laden and arguably more "artificial" processes—such as cloning or synthesization. There is a general perception that the ability to replicate artificially biological materials is very new; however, techniques such as fermentation and synthesis have relatively long histories. The arsenal of available techniques has, however, grown considerably over the past two decades as the biotechnological revolution has expanded the ways through which such processes of replication can be effected. Companies thus now find that they have at their disposal a number of potential methods for procuring biochemical or genetic materials: through in situ collecting or cultivation or through various types of laboratory-based manufacture.

While the idea of replicating rather than reproducing these materials has always been enticing, it has traditionally been complicated by a number of factors, principally the nature of the material itself. As the head of the NCI's natural-products development division, Gordon Cragg, explained, "some of these biochemical structures from natural materials are just so complex that you couldn't have dreamed them up if you tried, and this has made synthesis almost impossible as it would just involve too many steps, or at least too many steps to make it viable for large-scale production."[2] As Cragg suggests, until recently technologies were rarely sufficiently advanced to make replication of a complex agent more economically viable than simply extracting it from large quantities of naturally reproducing organisms—especially as there have traditionally been few constraints on collecting. Consequently, in the late 1980s, less than 12 of the 119 plant-derived drugs on the market were produced using replication techniques such as synthesis or plant-cell culture. Most were produced by extracting active compounds from whole plants grown in small-scale extractive reserves, broad-acre plantations, or greenhouses.[3] However, as Kloppenburg noted at the time with characteristic prescience, this situation was almost certain to change given the application that rapidly developing biotechnologies could have in the manufacture of "high-value, low-volume products like drugs and scents."[4]

Despite what might be imagined, even this reliance on plant cultivation did not always advantage source countries. For-profit companies have an obligation to their shareholders to reduce production costs and to seek out alternative sources of supply or means of production if they prove to be cheaper or easier to regulate. This has been illustrated several times in the recent history of natural products research and development, most notably in the cases of the Dioscorea yam and the rosy periwinkle. During the 1960s, the Dioscorea yam became the world's most commercially valuable pharmacological plant when it was discovered that it could provide a natural source of estrogen for the mass production of birth control pills. As Brush has reported elsewhere, Mexico was one of the world's largest producers of Dioscorea at this time, and the Mexican government formed a nationalized company to market the steroids, partly to ensure that some financial benefit be returned to the peasant collectors that supplied the raw materials.[5] However, when the Mexican government raised its prices, companies began to turn to other suppliers. Mexico's share of world production of the steroid consequently fell from 75 percent in 1963 to 25 percent in 1976.[6]

Later, in a precursor to what would become a more common practice, the companies began to turn away entirely from this process of collecting whole plants in foreign countries, preferring to concentrate their efforts on streamlining artificial methods of production. Francesca Grifo, formerly of the Fogarty International Center of the NIH and now director of the American Museum of Natural History's Center for Biodiversity and Conservation, explained in an interview how the largest pharmaceutical companies effectively trumped Mexico's attempt to increase their revenues from *Dioscorea* sales: "They [the pharmaceutical companies] just synthesized it and said ciao! . . . And that was it. Mexico kept asking for more and more, and that was just an incentive for the U.S. company to make it—to manufacture it."[7]

This case paralleled a similar one that had occurred in the 1950s. During this period, the drug company Eli Lilly discovered that two alkaloids extracted from the plant *Catharanthus roseus* could slow the progress of leukemia. The plant, known as the rosy periwinkle, was native to Madagascar and was initially produced there on plantations to provide a supply of the extracts for commercial production of the drugs vincristine and vinblastine.[8] Lilly continued to grow the plant there until "political problems forced them to transfer production to Texas."[9] Until recently, 1,000 tons of this plant material were grown annually in Texas for use in the production of the drugs. Although the drugs are reported to earn Lilly in excess of $100 million each year, no percentage of this profit is returned to Madagascar.[10] Whether any retrospective payment should be made to the supplying country in this case has been hotly contested. The material was, in fact, sourced from two countries—Madagascar and the Philippines, where it also grows— and of course it was collected many years prior to the introduction of any formal benefit-sharing agreements. It has historically proven both difficult and costly to extract and purify vinblastine and vincristine from whole plants, and it has been reported recently that the company is now moving towards the development of cell-culture systems in an effort to reduce the cost of producing the alkaloids while increasing yields.[11]

The political economy of biological resource production has, of course, undergone a profound change since the 1970s. Genetic and biochemical materials have become subordinated to capital through the introduction of new global regulatory regimes and are now constructed and used as alienable commodities. The era of free access to genetic resources ended in 1992, replaced by a new discourse of rights and responsibilities regarding the use of these materials, which directly affects both collectors and suppliers. Collectors (or at least the majority of them) are coming to understand that they must now provide adequate compensation for the use of genetic resources that form the bases of these and other pharmaceuticals, while suppliers are becoming cognizant of their right to capitalize on the genetic resources that have been declared to form part of their national patrimony. Moreover, as we have seen, protocols have been introduced that are designed to encourage their active and equal participation in the utilization of these resources.

Although many companies have now publicly acknowledged that countries and communities have a right to claim appropriate compensation for the use of their genetic resources, a number of corporate executives are privately of the view that the granting of these "possessory rights" has actually had a profoundly *detrimental* effect on in situ collecting. Constructing these resources as part of national patrimony has, they argue, only served to encourage supplying countries to view them as exceptionally valuable commodities for which substantial amounts of compensation must be paid. This, in turn, drives states to make what the companies consider to be "excessive" demands for payment, infrastructural development, or training in return for their use. These continued requests for consultation and compensation from supplying countries and the introduction of protocols requiring collectors to conform with these requirements, are, they suggest, acting as a serious deterrent to the continuance of in situ collecting. As one of Merck's senior researchers has noted, as far as the companies are concerned these increased demands quickly compromise the economic efficiency of their collecting operations:

> [We are] simply not willing to pay the excessively inflated prices that are being asked. So as a result there are a substantial number of countries where we simply do not collect anymore. Additionally, there are a lot of political hurdles going up, basically everywhere, stopping people from taking stuff out of the country, and then there are the places that will tell you "you build laboratories here to do the work, then we'll talk about you taking the materials out," and that adds a level of diseconomy I guess, that we are not prepared to enter into. *There are other sources of lead compounds.*[12]

"BUILD IT FOR US"

While securing ongoing supplies of biochemical compounds through in situ collecting has always been problematic, there have, until now, been few alternatives. As Bob Borris's last remark suggests, this situation is now changing. Technological developments are beginning to release corporations from their dependence on in situ collecting. As a direct consequence of recent advances in screening methodologies, synthesis, semisynthesis, and plant-cell culture, it is now possible for large corporations to take a collected sample of biological material and extract particular genetic materials or biochemical compounds of interest from it. These materials act as a proxy for whole organisms, and the proxy, in turn, acts as a kind of "foundational resource" that companies are able to exploit in a variety of novel ways. Companies are able to take these proxies and use them as templates to construct further quantities of the genetic material or biochemical compounds that they require, obviating the need to return to the original point of collection. Advances in synthesization are enabling a greater number of compounds to be reproduced in this way, while also reducing the amount of material required as a "starter" for such processes.

There are a number of distinct advantages to producing these materials through processes of artificial replication. One of the most significant is the ability to tightly control and standardize the production process, to reduce or eliminate all structural and environmental variables. Bob Borris explained in an interview why large companies like Merck favor this form of production:

> The fact that we can identify the compound, and we can now do that routinely *on a milligram or less of material,* basically opens the door for the synthetic chemist to come in and—if the activity that we are seeing is something that is sufficiently interesting—build it for us. And as soon as he [sic] can build it for us, we are no longer talking of having a couple of milligrams to work with, but having grams or kilos or whatever from a well-documented, repeatable procedure—which from nature is just not quite the same.[13]

In 1988, only 12 percent of all plant-derived drugs could be produced using processes of synthesis or semisynthesis. However, by 1996 both Steven King of Shaman Pharmaceuticals and Bob Borris at Merck were estimating that over 50 percent of them could be produced using these methods of manufacture.[14]

As Kloppenburg foresaw, similar advances have been made with another replication technique: plant-cell or -tissue culturing. Culturing allows a mass of undifferentiated cells called a "callus" to be maintained indefinitely in an artificial medium containing hormones (such as cytokinins) and organic substances. Tissue culture can be initiated using almost any part of the plant (or other organism, including humans), although some sources of tissue culture better than others. According to Kris Vencat, the CEO of Phyton Catalytic, one of the world's largest plant-cell-culturing facilities, replication techniques have advanced to a point that it is now possible to begin a process of plant-cell culture using a biopsy of less than four milligrams of material from a living, or even cryogenically stored biological sample. The advantages of the technique are that large numbers of cells may be grown in comparatively less space than that required to sustain populations of whole organisms. Moreover, the material can be gown in sterile conditions and in a controlled environment. The case of Taxol (paclitaxel) provides a salient example of the effect that advances in this technology will inevitably have on the traditional dynamics of supply and demand in natural products and, therefore, on processes of biological collection.

In 1962, when undertaking one of their earliest biological screening programs, the NCI discovered that extracts from the bark of the pacific yew tree, *Taxus brevifolia*, exhibited antitumor activity. It was not until five years later that researchers Wall and Wani first successfully isolated the active molecular compound, which was to be trademarked as Taxol.[15] They went on to fully elucidate the structure of the compound in 1971. The NCI undertook much of the preliminary research on the process of translating the compound into a safe and effective pharmaceutical. As this research was experimental in nature, only small amounts of the compound were required to conduct the initial tests. This was fortunate, as the bark from which the compound was extracted had to be stripped from a tree that is particularly slow-growing. The *brevifolia* can take up to one hundred years to reach a girth of some eight inches, and the process of stripping the bark will often kill the tree.

Although the NCI had invested a substantial amount of taxpayers' money in prosecuting this research, which culminated in a series of successful clinical trials, they were not in a position to bring the pharmaceutical to the market. As a public institution, they are prohibited both by law and under the terms of their operating grant from undertaking commercial production of any pharmaceutical or drug therapy. They were consequently obliged to seek out a corporate partner able to take paclitaxel through the final stages of clinical development and marketing for use as a proprietary pharmaceutical.[16] Bristol Meyers Squibb secured this contract in mid-1991.

Following a public announcement that the compound was showing remarkable success rates in the treatment of advanced ovarian and breast cancer, demand for paclitaxel—and consequently for the bark of the *brevifolia*—soared. It was estimated that some 320,000 kg of dried bark would be required to treat patients with ovarian cancer in the United States alone and that this would demand the sacrifice of some 35,000 trees.[17] Bristol Myers Squibb contracted Hauser Chemical Research to collect 1.4 million kg of bark over the following two-year period—a rate of extraction that was clearly unsustainable. The only alternative, at the time, was to conduct a worldwide search of related species of *Taxus* for any that might contain similar compounds and then attempt to convert these similar compounds to paclitaxel through semisynthesis. Compounds extracted from ornamental yew shrubs and hedges that were routinely cultivated in the United States and in Europe proved useful in this endeavor, and for a period of time the collection of *Taxus brevifolia* bark declined. Although this provided one method of supplementing production, it was evident that more efficient techniques would have to be found to meet long-term production needs.

The most attractive of these was to find a means of artificially replicating paclitaxel *itself*, rather than attempting to create analogues of it. This proved, however, to be an extremely difficult task. Although total synthesis of paclitaxel was achieved in the laboratories of Robert Holton at Florida State University in 1994, there was little confidence within the bioprospecting industry that commercial quantities could be produced using this or other replication techniques, such as plant-cell culture. When asked at the time to comment on the viability of either method, Steven King, the vice president of Shaman Pharmaceuticals provided the following assessment, which reflected the general consensus of opinion within the industry:

> Plant cell culture is not, in my view, going to be a really viable alternative. Synthesization . . . well, a lot of the compounds that are of interest are so complex—take Taxol for example. There are something like thirty-seven steps involved in synthesizing Taxol— and it's too complex to really make the whole thing viable. I mean it's been synthesized, but not for commercial production. So in the short term, no, I don't think it's really viable—but things are changing really rapidly—anything could happen in the next fifty years to change that.[18]

"Things," in fact, changed so rapidly that it took Bristol Myers Squibb not fifty years but only *two* years to realize their goal of producing commercial quantities of paclitaxel through plant-cell culture. They did this by drawing on the services of a specialist company, Phyton Catalytic. Phyton operates the world's largest dedicated plant-cell culture facility, a highly automated and computer-controlled laboratory that has a large scale fermentor with a capacity of some 75,000 liters. It was in this facility that they first began to produce paclitaxel (the named active ingredient of Taxol, now that Taxol is a registered trademark of BMS)[19] in 1995 using a new, exclusive, and patented plant-cell fermentation technology system.[20] The PCF process enables the plant cells to be grown in huge quantities in highly controlled specialized fermentor tanks in such a way that they produce paclitaxel at high rates. As Phyton noted, "by leveraging commercial-scale fermentation to unlock the plant cell's intrinsic ability to grow and synthesize compounds, PCF technology enables the sustainable production of plant-derived compounds without the need for any plant harvesting or chemical synthesis."[21] PCF is now reported to provide "the most advantageous method of manufacturing plant derived compounds . . . in a secure, sustainable and efficient way."[22] These, and other similar successes, clearly disprove arguments that plant-cell culture is not an economically viable means of producing complex natural biochemicals. As Gordon Cragg reflected in an interview that took place just one year after my interview with Steven King:

> Plant cell culture is a lot more difficult than synthesization, but it's been done with Taxol. Phyton—Bristol Myers's partner—are now producing Taxol from tissue culture, and I think this is a sign of the degree of success that they've had—that Bristol Myers is interested—because they [BMS] are not going to back something if they think it's going to cost them too much.[23]

By employing these technologies, it is becoming possible to rapidly generate viable quantities of some active compounds from very small amounts of collected material. The transition from reproduction to replication as the preferred means of production for biological materials is all but complete. While in the 1970s it was rarely considered more economically viable to replicate materials than to extract them from naturally reproducing organisms, this is no longer the case. Reporting on the relative contributions that biotechnology and bioprospecting now make to pharmaceutical development, *The Economist* has drawn attention to these changing dynamics, noting what many producers now know for a fact—that "although working out the growing conditions [in plant-cell culture] can be tricky, it is nowhere near as tough as trying to get tons more of an exotic plant from a far-flung corner of the earth . . . especially as plant samples can now be up to 100 times more costly than laboratory-based cell cultures."[24]

COMBINATIONS AND PERMUTATIONS

Collected samples of genetic and biochemical materials may be used as starters for these processes of replication, but this is not their only value to collectors. They may now also be modified and "combined," in a Latourian sense, to create entirely new collections of valuable genetic and biochemical materials, collections of bio-information. Castells argued that biotechnology should be understood as one of the new "informational technologies" on the basis of its ability to "decode and *re-program* the information embodied in living organisms."[25] He has also argued that the information age is characterized not simply by the centrality of information (as this is common to many ages of development) but more significantly by the rise of technologies that are able to *act on* information (by replicating, modifying, or transforming it) in order to produce new types of modified information that are themselves able to be commodified as resources. Some interesting examples of this can be found by looking at the ways in which biological materials and information are now utilized in drug development.

Much of the development of medicaments and pharmaceuticals over the ages has rested on an empirical strategy of testing the efficacy of plant or animal extracts against a range of disease conditions through some form of in vivo testing. This has traditionally involved a rather laborious "one on one" method, testing one set of agents against a single disease condition. In recent decades, as I noted earlier, new technologies and artifacts (the sixty tumor-cell-line panel, for example) have enabled researchers to conduct such tests in vitro, accelerating and broadening the rate of exposure of compound to condition. Technological advances—miniaturization in particular—have dramatically reduced the amount of reagent required to perform a test of efficacy, cutting the cost of screening while increasing the number of screens that may be conducted at one time. These improvements have greatly increased the demand for novel molecular compounds. This, in turn, created a supply problem. Many of the molecular compounds that had been derived from natural materials had already been screened and were yielding only biologically active molecules that were already known. Scientists recognized that one way of increasing the total number of potential drug candidates

would be to create entirely new classes of compounds by combining existing small organic molecules known to have some proven action or efficacy
against disease conditions. This goal was achieved in the mid-1990s by the
development of a new technique—combinatorial chemistry.

Put simply, combinatorial chemistry involves creating large collections,
or "libraries," of medicinal compounds by rapidly and systematically synthesizing all possible combinations of existing smaller chemical structures
or building blocks—such as nucleic acids, peptides, and small organic molecules. Traditionally, an organic chemist would combine two such elements
to produce a third, which would then be tested for efficacy. Increasing the
number of elements combined in each part of the equation produces a significant multiplicative effect. So, where A + B would normally produce C,
if A is already a combination of five of these elements, and B is a combination of another ten, then C will automatically produce fifty distinct compounds.[26] This results in an exponential increase in the number of novel
compounds available to be screened for activity against various ailments.

In the past, the technology to carry out this process simply did not exist,
but, in 1994, the company Affymax, now a wholly owned subsidiary of the
Glaxo Wellcome corporation, built one of the first pieces of equipment dedicated for this purpose. The "encoded synthetic library synthesizer"—as it is
known—employs computer software, chemistry, and automated technology
to enable hundreds of thousands of new compounds to be created and
screened each week. In December 1999, Sir Richard Sykes, then CEO of
Glaxo, reported that in a two-week period in 1998, Affymax had synthesized
1.67 million compounds—a number in the same order of magnitude as the
total number of compounds synthesized by the entire pharmaceutical sector prior to 1990.[27] These compounds, which are produced by combining
existing organic *and* synthetic substances, are preferred to naturally derived
compounds for a number of reasons. They are often more efficacious, as
they are a hybrid that combines the best features of different parent molecules; they may be reproduced in the laboratory under controlled conditions; and they are engineered and, hence, readily patentable.

To produce a large number of compounds is one thing—to accurately
assess their efficacy is another. In order to keep up with the vast number of
new compounds being generated, companies began to develop parallel
screening protocols. These enabled hundreds of compounds to be evaluated simultaneously using rapid, high-throughput, predictive assays. The
assays are used to assess factors such as the potency of the compounds,
their selectivity for the target, how they are absorbed and metabolized,

their tissue penetration, and their potential carcinogenicity. One of the principal objectives of this research process is to establish, and indeed predict, which particular combinations of molecules are likely to show greatest activity against specific disease conditions. By employing this technique, scientists are able to continually refine the process of recombination, so that the most efficacious elements of compounds are identified, isolated, and recombined to create smaller and smaller libraries of substances with highly specialized actions.

This highlights another of the ways in which companies utilize collected biological materials. In thinking about how to improve the efficacy of any possible compound, researchers are concerned with obtaining information about the molecular structure of the pharmacologically active compounds contained within natural materials. This biochemical information is valuable, as it may be used in this process of refining the manufacture of engineered compounds. It is important to recognize that this information may become a key commodity *in its own right*, able to be extracted, notated, stored on databases, acted upon, and utilized quite independently of the biological material from which it was drawn. The information alone may have a productive value, without ever being realized in a material form. For example, researchers are now employing this derived biochemical information to create what might be termed "virtual compounds." Rather than attempting to build new molecules "for real," in the laboratory, scientists are instead taking information about the structure and actions of existing compounds that they have drawn from analyses of collected biological materials and feeding it into computer programs similar to those used to forecast weather patterns in order to predict the properties of prospective compounds *without actually making them*.[28] They may take information about the structure and action of an existing compound (for example, one that looks promising but binds weakly to a target) and use that information to model, computationally, how to amplify that effect in an engineered analogue.

It would be impossible to undertake these complex computational tasks without the development of new informational technologies. It would be equally impossible to do if the genetic and biochemical information could not be rendered as a form of digitized information able to be read and manipulated by these new informational technologies. Biotechnology and informational technologies have here combined to create a new industry, "bio-informatics," and another set of commodities—highly valuable collections of biologically derived information stored on databases. Collectors, it could be argued, are able to employ both biotechnologies and informational technologies to decode and *reprogram* or recombine both the substances and information embodied in living organisms in order to produce new collections of material that are commodifiable and that thus provide them with further sources of economic productivity, in a kind of cumulative feedback loop.

These technologies have also allowed researchers to focus the search for active or engineered compounds to such an extent that that they waste very little time pursuing unpromising leads. The implications that these developments have for improving the efficiency of drug identification and development processes are difficult to underestimate. Glaxo's UK research director, Clive Dix, recently highlighted the dramatic improvements that the introduction of this directed, informational approach has had on research and development turnover times at Glaxo: "Large combinatorial chemistry libraries were screened and then smaller focused libraries generated around the most interesting 'hits.' 120 compounds were then evaluated in parallel assays, six of them met the project criteria and one has been selected to take further. All in all this took 40 chemist weeks and four biologist weeks, traditionally it would have been some 660 scientist weeks."[29]

Various executives involved in collecting and drug development have described the potential of combinatorial chemistry and speculated about the impact that it may have on collecting operations. Rob Thomas, a biochemist and executive director of Biotics, one of Britain's largest private bioprospecting firms, explains why this new technique is so important for researchers:

> Well, a lot of good-selling drugs are modified peptides. In the past, peptide synthesis was a very tedious and specialized business, but now it is automated—you can virtually dial it into your computer, and it will stick these things together [and] come up with what you want. And as a consequence of that, it is possible to throw in a bunch of starting-unit amino acids and instruct it to synthesize them in a sequence. But you can also throw in a mixture and therefore it will synthesize the mixture, so if you throw in ten amino acids, you can end up with thousands of compounds . . . all in the mixture. That is, of course, pretty useless unless you can separate them all. But there are ways of doing this so you can extract discreet combinations of compounds. You can tag them so then you can backtrack if you find ones that are active and find out which one was the actual compound. So you are actually now synthesizing

thousands, tens of thousands of compounds at the same time—so you can see why all these companies are getting into combinatorial synthesis."[30]

As Gordon Cragg of the NCI notes, combinatorial chemistry is also significant in providing scientists with the capacity to rapidly generate types of biomolecular compounds that exceed those currently found in nature. The resulting collections of biochemical compounds are examples of what Latour and Serres refer to as "boundary objects"[31]—entities that are a fusion of the organic and the technical:

> A lot of them are already going to this new combinatorial chemistry—where you take a particular molecule and you throw in a whole bunch of different reagents which will react with the molecules in different ways, and you get an incredible mixture out of it, and then you test that mixture for any activity, and then you try and isolate the particular group of molecules responsible for that activity. . . . They now know what the genes are that are responsible for particular enzymes so now they can actually do what they call "mix and match" and come up with molecules that the organism doesn't make itself![32]

Within the pharmaceutical industry, competitive advantage increasingly rests on the ability to secure access to a host of novel "lead" compounds that could form the basis of new drugs. Most companies now prefer to acquire these lead compounds by accessing either existing or specially generated "libraries" of DNA, RNA, peptides, or small organic molecules. Many of these libraries are made up either entirely, or in part of molecular compounds or genetic materials extracted from organisms collected in developing countries on bioprospecting expeditions. Proxies such as cryogenically stored tissue samples, cell lines, molecular compounds, and the like have played an important role in the facilitating the mobilization of these "key" or "essential" elements of biological materials—allowing functional genetic and biochemical components to be collected and transported from one geographic location with much greater ease.

Proxies have also played an important role in facilitating the mobilization of these key elements on a much more "domestic" scale. These proxies act as a kind of immortal, artificial body: an environmentally controlled, long-term, and secure bank of exploitable molecular and genetic resources. With the development of techniques such as combinatorial chemistry and gene splicing, it has become possible for scientists to extract viable fragments of genetic and biochemical material from these proxies and combine them in ways and at speeds that were previously inconceivable—and all within the safe confines of the laboratory. The number of ways in which collected and extracted genetic and biochemical materials and information are being combined is almost limitless—multiple thousands of combinations are created in laboratories every day. As Thomas noted, scientists are having to employ complex systems of florescent tagging of different reagents in order to trace particular compounds as they pass through ever more baroque cycles of combination and transmutation. It is important to remember, however, that his task is being performed with the intention of tracing active agents so that the capitalization they have acquired through the process of recombination is not lost. It is *not* being performed with a view to establishing or recording the provenance of those components.

The act of extracting and engineering genetic or biochemical components seems to produce a peculiar disassociative effect. The libraries of viruses, molecular compounds, cell lines, and the like that are produced out of collected materials are no longer understood or constructed as part of the patrimony of a nation state but rather as technological artifacts that remain the property of the organization or corporation that developed them. With each stage of processing—at each remove—awareness of the provenance of the materials and of the obligation to compensate for their use diminishes a little further. For the scientists that use these artifacts (those involved in combinatorial chemistry or genetic engineering), the derivation or provenance of the material is, understandably, largely irrelevant. After all, in the case of combinatorial chemistry, the compound generated may contain only one or two small molecules from a natural product among many others that have been synthetically produced. And yet, we can think of those few molecules as being just like the five signature notes of the "Live and Let Die" title song that have been so frequently sampled for inclusion in television car advertisements and new rap songs. Although forming a relatively tiny part of the whole composition, they may well prove to be crucial to the cohesion and success of the new work.

Scientists are relying more and more on securing access to existing, or especially engineered collections or libraries of genetic and biochemical materials, using them as an essential resource for undertaking basic research. Despite this, here has been little discussion or investigation of the effect that this development might have on altering the existing dynamics of trade in natural products. The introduction of new techniques such as cell culture and combinatorial chemistry will inevitably serve to revalue existing collections by expanding the number of ways and times they may be used or reused. What implications might this have for the whole project of in situ collecting? In answering this question it is important to begin by considering what incentives there are for organizations or corporations involved in natural-products development to continue with in situ collecting and why they might, alternatively, prefer to adopt other methods of procuring and reproducing interesting lead compounds.

THE DIMINISHING ROLE OF IN SITU COLLECTING

There is little doubt that these technological developments will have a dramatic impact on the social and spatial dynamics of trade in natural products. As Gordon Cragg suggested earlier, the ability to secure biochemical compounds through processes of replication or recombination seriously undermines the viability of existing collecting projects. While this is true, it is unlikely, in my view, that the introduction of these new biotechnologies will result in a complete cessation of collecting. Rather, they are more likely to induce a subtle transformation in the *dynamics* of collection—notably in what is collected and when. Pharmaceutical companies may well continue to undertake some in situ collecting; however, this collecting process will be much more tightly focused. They will almost certainly abandon the large-scale collection of bulk quantities of natural materials and concentrate instead on acquiring only those organisms that they feel might yield patentable "lead" compounds. These compounds will then be engineered or synthesized in the laboratory to provide the quantities of material necessary for large-scale production of the drug. The laborious and costly task of returning repeatedly to source countries to secure further supplies of the biological materials from which those compounds are drawn will not continue—not when new technologies provide more secure methods of production. As Neil Belson the president of Pharmacognetics, a Maryland-based biological collection firm, confirmed, there are important reasons why large corporations prefer to replicate these materials in the laboratory rather than gather them in the field:

> There's lots of reasons why they'd prefer to synthesize these things—it's easier to synthesize things, and it's a lot cheaper than having to go back to a natural source; it's a lot more reliable as a supply source, it's a lot easier to standardize production because activity levels might vary widely from plant to plant. They really just

want to use natural materials as leads—to find the structure. . . .
Then they will just go from there and synthesize a hundred thou-
sand variants of it that might be even more effective and less toxic.[33]

There is also an important temporal advantage to moving toward
laboratory-based replication of genetic materials and biochemical compounds. As I have noted here, advances in cryogenic storage allow samples
of biological material to be stored in such a way that their genetic and biochemical components retain their viability over long periods of time—for
hundreds of years at least. This allows collectors to investigate, manipulate, and replicate these resources at their leisure. The broader economic
or political environment may change, new legislation or policies may restrict the types of collecting projects that can be undertaken in the future,
species may decline or even become extinct. None of these events need
necessarily concern collectors greatly: their cache of materials remains intact, archived, and available for long-term future use.

It is already evident that companies are feeling constrained by the need to
negotiate complex agreements to access bulk quantities of biological materials and that they are disinclined to pursue this course of action when it is now
possible to replicate the genetic and biochemical materials that they require
within a closely controlled laboratory environment, well insulated from climatic or political variables. As Steven King, one of the most vocal supporters
of bioprospecting, admitted frankly in interview, "a lot of pharmaceutical
companies are not interested in these compounds if they are restricted to having to get a supply from raw materials because of the perceived noncontrol of
the source and the exposure and liability that is thought to be associated with
that."[34] Although collecting agencies such as the NCI and the NYBG have
made, as part of their stated commitment to the equitable and sustainable
exploitation of genetic resources, an explicit commitment to "seek as a first
source of supply natural products from the source country," their commercial
partners have been less willing to obligate themselves in this way. As a consequence, most contractual agreements between collectors and companies
contain an escape clause that allows companies to opt out of their commitment to return to the source country for further supplies of materials, if the
source country "cannot provide adequate amounts of raw materials at a mutually agreed fair price." Suppliers of genetic and biochemical resources are,
of course, just as vulnerable to the pressures of "global sourcing" as producers of any other raw materials. These resources are derived from organisms
that often grow in several countries, and companies are not averse to playing

off one supplier against another in order to reduce costs. The threat of losing a contract to a neighboring country may prove sufficient to temper any potentially "extortionate" demands, as Gordon Cragg has noted:

> In our "Letter of Collection," we say that we will certainly ask the company as part of the licensing agreement to use as its first source of raw material the country where the plant was first collected, but we do make the point [to the supplying country] that we can't hold them to that. If they get a lot of trouble from that country and next door the plant grows equally well—if this country is really asking an arm and a leg for the stuff—then we really can't stop [companies] from going next door. This is what companies get worried about— that countries will come out with outrageous demands.[35]

The question of what constitutes an "outrageous demand" is an interesting one. In accordance with the new protocols set out under the biodiversity convention, many nations began to establish minimum terms and conditions that interested companies would have to meet in order to secure permission to begin collecting within their borders. Most, although not all, required companies to sign contractual agreements that included a commitment to provide compensation in three phases. The first compensatory phase is not particularly onerous for companies. Some nations required them to pay an up-front payment or fee for a collecting permit; others insisted that companies agree to provide equipment or infrastructure to help them establish their own drug-development facilities. However, companies were unused to paying for access to genetic or biochemical materials, and many struggled greatly to accept the legitimacy of these new demands.

As I noted earlier, many came to view these as "additional" costs, arguing that they were "excessive" and "over-inflated." It would be a fortunate business manger indeed who found it possible to secure all his or her industrial raw materials for free, and yet this seems to have been the expectation that many executives in the bioprospecting industry continued to hold in the early 1990s. While they later began to accept the need to pay for the use of these resources, they remained unsettled by the notion of "compensation," which, it seemed to many, emanated from a need to redress a broader, generalized postcolonial malaise—a lingering and unresolved sense of injury, loss, and dispossession. It is certainly true that many suppliers in developing countries did feel mistrustful of these new collecting enterprises and they were determined to ensure that, on this occasion, contractual negotiations would proceed on their terms.

Although the negotiations for the establishment of new contractual agreements governing the collection and use of genetic and biochemical resources were initially conducted in a spirit of détente, they were undoubtedly also imbued with these subtle and complex tensions. What resulted was the generation of "compensatory" agreements rather than the establishment of a transparent commodity market in genetic materials. The dissimulation created confusion about countries' demands and expectations, as well as companies' obligations. Companies believed that payments should reflect only the market value of the goods currently being transacted (although that was clearly impossible to predict), while many supplying countries felt that payments should provide this and perhaps something further, some compensation to "reduce or balance" the injuries and losses associated with past injustices. The turmoil that ensued was described with chilling accuracy by a senior executive at Bristol Meyers Squib:

> Some of the countries wanted large amounts of money on a per-sample basis; they wanted the pharmaceutical company to go there and teach them to how to do the extractions or whatever on site — before any positive indication that they had anything worthwhile coming out, and, in essence, that it is not economically viable for us. To say if a certain number are successful, or if one is, then as part of the compensation for us to establish some sort of laboratory there . . . *maybe*. But another country wanted all of our excess equipment and it was a *huge* task just trying to determine what we had in the store house, how this stuff could be sent, how it could be imported, and it just ended up in a big jumble. They couldn't tell us how it could be shipped, how it would be stored, how it would be imported, whatever. So an awful lot of expectations and demands without knowing if they are feasible, even from their perspective.[36]

The perceived diseconomies of in situ collecting—"inflated expectations," demands for infrastructural investment, and political instabilities in source countries—quickly combined to make this an increasingly unattractive proposition for many large companies. Ironically, although bioprospecting was initially mooted as a means to conserve endangered environments, the argument that large-scale, in situ collecting was both environmentally and economically *unsustainable* was also used by companies to legitimize the move to laboratory-based production. These factors, combined with a desire to secure greater control over the production process, provided powerful incentives for corporations to move to artificial methods of replication, such as plant cell culture, as this pharmaceutical company researcher suggests:

> Plant-cell culture to me is a way of producing plant constituents in an environmentally neutral fashion. From the standpoint of a pharmaceutical company, for instance, if my work were to generate a compound—from, well, let's say not an endangered species but [one] with a very limited distribution, there would be a lot of pressure on me within the company to come up with an alternative sourcing mechanism because, number one, it is against our corporate policy to do any major damage to any population of a species in the environment. So if I wanted to make fifty kilos of compound that would require fifty trees and there are only fifty-five of them left, guess what? We're not going to do that. Alternatively, if we were to come up with a plant in some hypothetical Third World country, where the distribution was limited to that country, and the government which was formerly friendly suddenly goes through some changes and becomes hostile, so cutting off our supplies, then it would be catastrophic for us. So we have had to look at agriculture as a possibility, but, liking to have everything well under control in a scientific sense, well, if we can induce it in culture then it would be a much, much more straightforward way of producing it.[37]

The implications this has for the whole project of collecting are apparent and were clearly articulated in another interview with a senior executive at Bristol Myers Squibb:

> There are no guarantees [with cell culture], but we've made a lot of progress. One can envision perhaps doing *an initial collection* and storing some plant material, and then when you did have your hit—it's an arduous process as well—establishing a cell line and culture and trying to tease out the compounds of your choice. But once this art is developed, great, now it is a fairly crude science, but it will become an art, then a science, and to my thinking this will be the way of natural products in the future.[38]

The likelihood of companies returning to source countries for *ongoing* supplies of genetic or biochemical resources is rapidly dwindling. Only compounds that prove too complex to replicate via existing technology will be resupplied from wild collection. As technology advances further, these will inevitably fall in number. These developments cast serious doubts on the claim that bioprospecting is likely to contribute to the economic development of supplying countries through the development of cash crops or extractive reserves. It is clear from my interviews that pharmaceutical companies do not have, and indeed may never have had, any intention of securing commercial quantities of genetic or biochemical materials from either national parks or extractive reserves in source countries. They, at least, seem quite clear about how such material will be produced now and in the future. As one executive frankly stated, "we set up on the assumption that anything that we find will eventually be made by total synthesis—we're not really looking towards producing material from wild specimen collection."[39] Another concurred: "As far as sourcing goes, you're absolutely right. . . . I don't think the answer, the best answer, is to go back and even cultivate it, let alone collect it in the country of origin. My particular flavor of what the right answer is plant-cell culture."[40] As I have illustrated here, many large corporations and organizations have devoted considerable resources over the past decade to undertaking extensive collection programs. Having amassed large collections of genetic and biochemical materials and transported them back to their laboratories in the United States, they have little need to institute further operations, preferring instead to concentrate on exploiting the materials that they have already acquired often under extremely favorable terms and conditions. Steven King confirmed this when asked to reflect on the motivations behind the exponential increase in collecting that has occurred in the last two decades since the biotechnological revolution began:

> There does seem to have been a major global grab for genetic resources by major pharmaceutical companies and other corporations—I mean we [Shaman] don't think that way—but that's what most of them are doing right now. Bristol Myers Squibb contacted all sixty-four applicants for the ICBGs and asked them if they wanted to go in with them because they want to stockpile as much genetic diversity, biological plant diversity and probably other organisms—marine, microbial, etc.—as they can, and then put them into repositories so their long-term corporate interests are safeguarded. Then they can just keep bringing new molecules on-line, and there is no question in my mind that that is going on, that they are following that path and while everyone is whizzing around trying to figure out how to regulate it, they are just stockpiling it, and by the time they [the regulators] have gotten it all figured out it

won't matter because these corporations will have gotten some *huge* quantities of things stockpiled and it will be too late to do any sort of retroactive regulation.[41]

Heavily involved in bioprospecting for the last decade, Bob Borris at Merck has confirmed that the end to collecting may now be in sight: "Personally, I see the current heyday of collecting ending by the turn of the century [twentieth]. The politics involved in this will get severe enough that we may wind up collecting in just two or three countries. It's unfortunate, but it's an economic situation. . . . We may just retract to North America and spend our dollars here."[42]

In the event of companies abandoning in situ collecting as a means of sourcing ongoing supplies of genetic and biochemical material, the only way in which bioprospecting can possibly make any long-term contribution to economic development in source countries is in circumstances where a negotiated royalty or fee is forthcoming for the successive uses that are made of the materials already collected under these programs. Equally, the only contribution that it can possibly make to conserving endangered environments in source countries is in instances when a percentage of this return is committed, and actually put to use, for conservation purposes. The probability of either of these occurring is discussed in greater detail in the following chapter. Although companies are inexorably moving toward sourcing genetic and biochemical materials by artificially replicating them using techniques such as synthesis and plant-cell culture, there has been scant acknowledgement of this fact within the bioprospecting literature. This consequently raises a question about the degree to which this eventuality has been anticipated and is reflected within the contractual agreements that govern the terms of use for the collected material—a matter that is also discussed in greater detail in the following chapter.

THE ADVENT OF MICROSOURCING

The possible curtailment of in situ collecting should serve to concentrate attention even more sharply on the fate of materials that have already been collected and the uses to which they may be put, both now and into the future. For while some companies *may* continue to secure lead compounds through in situ collecting, many are clearly finding it more expedient to create them using techniques such as combinatorial chemistry or to extract them from samples of material that are more accessible—both geographically and politically. Rather than relying on securing supplies of genetic or biochemical materials through the traditional technique of "global sourcing," evidence suggests that many companies are finding that they now have another viable alternative: to locate an ex situ sample of the same material and to use that as a starter for processes of replication or recombination. I refer to this practice as "microsourcing." With so much of the latent potential of existing collections now able to be explored and realized in new ways, it is hardly surprising to find that companies are now more interested in "bioprospecting" existing collections than they are in undertaking new collections. Many companies in the U.S. pharmaceutical industry are in the process of making a transition from global sourcing to microsourcing as their preferred means of securing access to genetic and biochemical materials—a transition that marks a turning point in the history of natural-resource supply and economics.

There are a growing number of companies in the United States and around the world that specialize in supplying companies with genetic or biochemical materials that have been drawn from existing collections. Neil Belson, the CEO of Pharmacognetics, has confirmed that companies are more interested in the viability of lead compounds than they are in their provenance:

> If you get a compound that works, pharmaceutical companies won't care where it comes from. It's just the question of where they are going to look to find those compounds! But if you can tell them, "Hey I've got a good compound and I've got data to back it up," they're not going to care where you got it from as long as it works and it's safe. Combinatorial chemistry and natural products research: the only difference is where they are going to look for those lead compounds. But once they've found them, they'll buy them from anywhere.[43]

There are two types of existing collections that are of particular interest to bioprospectors. The first is collections of biochemical compounds or extracts that can be reexamined using new technologies and recombined with other synthetic compounds in a myriad of ways to form new products using combinatorial chemistry. Many such collections have been created over the years by university departments and research organizations but have languished in the absence of any new technologies of exploration or exploitation. The development of such technologies in recent years is having the effect of revaluing these existing collections while further reducing the incentive to continue in situ collecting. As the head of natural-products research at the NCI recently explained, he is under increasing pressure to draw upon their existing repository of collected material in order to create novel compounds or enzymes using combinatorial chemistry, rather than continuing with the more risky process of negotiating and undertaking further bioprospecting expeditions in what has become a rather fraught political climate:

> Our new director is very much into this area—he thinks that going out and scouring the world for plants and marine organisms is not exactly science. He says, "Look, we've got these enzymes, we've got these genes now and we can introduce these genes into a host organism—just a certain select group of genes, we don't have to put the whole lot in—and see what we come up with. . . . The thing is, is that combinatorial chemistry is quite a feasible process, and they've got total control of that process—they don't have to go to country x for more, and already companies are getting into that and pushing natural products to one side, wrongly I think, because combinatorial chemistry is only as good as the molecules you start off with and in nature you are always going to find something new. It is wonderful, but we are just at the start of that now, so I would say let's not rush into throwing the rest of it [in situ collecting] out.[44]

The second type of existing collections that are of interest to companies are collections of plants or organisms that may provide viable sources of replicable genetic or biochemical materials. Until very recently, this meant exclusively *living* collections. However, extraordinary developments in DNA extraction techniques are dramatically extending the number and type of collections that may prove useful to bioprospectors interested in securing supplies of such material. A growing number of museums of natural history—such as the American Museum of Natural History and the Smithsonian—are now beginning to store collections of tissue samples cryogenically. They use them for purposes of taxonomic identification and molecular and genetic investigation. However, as John Kress, senior curator of botany at the Smithsonian, confirmed, the materials may also be used as a supply of replicable genetic and biochemical materials and may, consequently, prove of interest to a much wider range of consumers:

> The cryogenic samples, the frozen tissues that are down to minus 360° C, we store in long-term storage facilities, and they could be

kept forever as DNA samples—not that we intend to clone them to re-create the plants, although you could. But they are essentially DNA banks that we are using to understand the relationships of plants and evolution. We've got three freezers full of those now, and out at the molecular lab the Smithsonian has developed a whole archival system of storing specimens actually in liquid nitrogen, not in freezers that can always break down. So it's kind of a long-term project. Whether they will last 400 years or not like the herbarium specimens, we'll have to wait and see, and as far as the living specimens are concerned, well, we have greenhouses to store them in. Not all of them are related to the bioprospecting initiative but some of them will be.[45]

Moreover, as Henry Shands, Head of the USDA's Genetic Resources Division suggests, developments in DNA-extraction techniques are advancing so rapidly that it is now even possible to use dried herbarium specimens as sources of replicable DNA. This development has created consternation amongst the holders of scientific and academic collections. Realizing, often too late, that their collections have been illicitly accessed and used as sources of industrial raw material, some curators now feel compelled to close access to their collections:

> Extracting DNA from herbarium samples is an absolute possibility now and the systematics people are worried because all of a sudden the doors to exchanging these collections are being closed.[46] Sharing around the world has been very good until recently, when there have been cases where people have been able to take the DNA— that is, extract DNA from herbarium specimens to the disadvantage of whomever is holding that specimen—and determine the gene sequence or the unique properties associated with that particular plant. On the pharmaceutical side of things, you can create an analogue that is related to some chemical that was in it, or, in the case of other plants, there is the possibility that they have looked at pieces of the code that would relate to the chemical structure. It is viable and it is happening and the reaction of everyone that is holding this material is, "Hey, I don't want that to happen to me," and so for now the doors are closing on exchanging these materials. I suspect that material transfer agreements of a very specific nature will evolve out of this to ensure that everything is kept on a very business-like basis.[47]

The collections of biochemical compounds, living or cryogenically stored materials, and even herbarium specimens that have been created by systematists working in institutions such as universities, museums, or botanical gardens are of particular interest to companies, as they are more likely to be have been accurately identified and classified. They are also less likely to have been patented, as they have been investigated for purely academic purposes in publicly funded institutions. This is a particularly important consideration for companies whose intention is to use the material for the manufacture of proprietary products. Material that has already been patented — such as those collections of engineered cell lines or organisms archived in patent repositories — are consequently of little interest, being only useful as production "tools."

That existing collections of materials can now be used in these ways is creating a new market economy for samples of genetic and biochemical material drawn from existing collections, and new opportunities for trade in the same. What are the dynamics of this nascent trade in samples of genetic and biochemical materials? Who is buying and who is selling in this new resource economy, and how does it operate and to whose benefit?

RE-MINING EX SITU COLLECTIONS

Any process of trade must begin with demand. Why is it more advantageous to source material by accessing existing collections rather than by undertaking a new collection program? The advantages of accessing existing collections are political as well as technological and economic. As a consequence of a loophole in the biodiversity convention, materials held in ex situ collections accumulated prior to 1992 are not subject to the same protocols as material collected in situ in source countries. Those able to access collected samples of material held in ex situ collections are under no obligation to compensate the original supplier of that material for any current or successive use that they might make of those materials, and the holders of such collections are under no obligation to require them to. Pharmaceutical corporations generally have a large enough research and development budget to be able to absorb the costs associated with paying compensation for the use of genetic resources. Moreover, most have acquired enough material over the last two decades to ensure that they are "self-sufficient," although, as one senior corporate research chemist has suggested, there are occasions when it "increases your odds to access say, focused collections of particular taxonomic groups" such as those held by academic laboratories.[48] Identifying or accessing such a collection would be a simple process:

> If you went back through the literature for the last fifty years or so and saw who was doing natural-products research, it would be relatively straightforward to generate a list of fifty to one hundred people, contact them directly regarding access to their sample collections and negotiate a fair price. . . . There are libraries of natural products that have been collected and isolated by various academic groups, and you can access those if you are willing to pay the right money.[49]

Smaller biotechnological or pharmaceutical companies, however, find the expense, time commitment, and complexity of undertaking expeditions in source countries and complying with the regulations governing access and use prohibitive. Many are also unwilling or unable to meet proposed levels of compensation. For these companies, it is simply more expedient and economically viable to microsource material from existing collections:

> Obviously ex situ collections are a way to get around the treaty [biodiversity convention], . . . and obviously it makes it easier because you don't have to negotiate access permits, et cetera. Because even if there is no real disagreement on the terms of the agreement, the mechanics—from initial proposal to signing the final agreement—are so complex and so uncertain or unsettled that well, you know—you're not even certain that you are going to *have* a collection in a year or two, and I mean if you're not even going to have a collection a year after you started—well how are you going to justify that to whoever does the budget for the company?[50]

This raises the question of which existing collections are being accessed in this way and of how collected materials are being traded or exchanged. As I illustrated in the previous chapter, the range of actors involved in commercial bioprospecting has changed quite dramatically over the last decade. Some, such as academic collectors, researchers working in publicly funded institutions, and what might be termed "corporate collectors," have a long history of involvement,while others, such as conservation agencies, are new to the game. The connections and linkages—the interdependencies in fact, between these different actors have thickened tremendously in the last decade as they have become more cohesively networked through new public/private bioprospecting partnerships. Despite this, these groups continue to be characterized and presented in the bioprospecting literature as sepa-rate actors that have, as Kloppenburg and Balick, have suggested, "different interests and objectives" in relationship to the collected materials.[51] Within the model that they put forward, all corporations and companies are grouped together, presumably by virtue of their shared interest in the com-mercialization of the collected resources. "Nongovernmental" groups, such as the NYBG, the Nature Conservancy and Conservation International, are placed into a separate category presumably by virtue of their shared interest in preserving collected material for a variety of other noncommercial uses. Government agencies such as the NCI and the USDA are placed into yet another category as trustees of collected materials. Each individual agency is also categorized as being "*either* a supplier (donor) *or* a recipient (deman-der) of germplasm."[52]

While it is not my intention to argue that these actors are all of the same ilk, I do want to challenge the presumption that they are completely distinct from one another, either organizationally or in terms of their interest in the collected material. Contrary to what is implied in Kloppenburg and Balick's model, it is no longer only pharmaceutical companies that are interested in exploiting their collections of material for commercial purposes. Evidence from this research suggests that many publicly funded organizations and in-stitutions have become actively involved in trading collected samples of ge-netic and biochemical materials. Many are allowing commercial interests direct access to biological materials in their existing collections for a fee. Others have agreed to allow companies to underwrite the cost of one of the institution's existing collection programs on the understanding that the in-stitution will collect particular specimens on the company's behalf while in the field. Yet others have agreed to effectively "rent out" their samples of col-lected material to companies for investigation and possible replication.

The ethical implications of "re-mining" existing scientific or academic collections are particularly disturbing. Foremost among these is the fact that material collected for "scientific" purposes is now being utilized as a com-mercial resource without the consent of the supplying parties (which may now be impossible to obtain). In many cases these activities are not in the public domain and are therefore not subject to public scrutiny. Some senior executives in academic or scientific collecting institutions are beginning to express considerable unease about the ethical implications of these activities:

> I shouldn't even really mention the name of the botanical garden, because I hope they're not doing it, but I had a long talk with them

about exactly that. A pharmaceutical company approached them and said, "Can we bioprospect your living collection?" And I don't know whether they said yes or no, but I personally have a real problem with them saying yes, because those things were collected for a different purpose. Maybe some of them were collected a way long time ago under no purpose at all—you know, just colonialism, you go and find some cool plants and bring them back and you don't have to ask the natives—that was acceptable then, that was what people did, but the fact is that to then take those collections and to reap a profit from them and to have *no way* to return a benefit to them . . . you know, I have a real problem with that.[53]

Although the risk of adverse publicity usually deters the largest pharmaceutical companies from bioprospecting ex situ academic or scientific collections, such practices are not unheard of. Pfizer, one of America's largest pharmaceutical companies, launched a new natural-products drug-development program in 1993. In announcing the program, Pfizer reported that it would not need to implement the requirements for compensation devised under the convention, as it was intending to bioprospect exclusively within the U.S. mainland.[54] However, in the course of my research it was bought to my attention that Pfizer had arranged to collect samples of material in the Fairchild Tropical Botanical Garden in Florida. These specimens were originally supplied to the garden from various developing countries in South and Latin America in the interests of broadening public exposure to and understanding of Latin American flora. Interestingly, Pfizer subcontracted field collectors from the NYBG to undertake the collection on their behalf. One of these field collectors described to me the way in which this process took place:

> The collecting program that I went on—we went to down to Florida to Fairchild Tropical Garden at Coral Gables, and we collected tons of species there that had been imported and planted out and cultivated. We were shipping them back to the garden [NYBG] then breaking them down to create an alcohol extract and then shipping that back to Pfizer where they start doing the screening. . . . If they got a hit they would try and find enough in the States; . . . if they couldn't get enough from somewhere like Fairchild they would call a seed house that could raise them in a hot house and just order them. Towards the end of the project

when we still had a few really hard-to-get plants that either weren't native or they were tough to grow or they didn't really occur in the wild anymore, well we just went and ordered them from the hothouse so we could get bulk samples.[55]

There is nothing illegal about either Pfizer's or the NYBG's activities, although some would find them questionable. Organizations such as the NYBG that have commercial contracts with Pfizer and scientific contacts with Fairchild play a pivotal role in facilitating this new trade by creating a vital link between the commercial interest (Pfizer) and the botanical garden (Fairchild) while at the same time obscuring the existence of this link from public view.

> You have to give them [Fairchild] a written description of why you want the sample and the information and that has to be approved. . . . I think that they [Pfizer] do it through the garden [NYBG] because herbariums and botanical gardens recognize each other and they lend to each other all the time, so Pfizer would rather do it that way. . . . We get a permit because the NYBG recommends us—and they [the NYBG] give the Fairchild a certain monetary payment like 250 dollars per visit and what we do is we run around all day locating plants that we haven't collected yet and we take living samples from them—we take cuttings—one specimen for the NYBG and one for the collection that we were making just for Pfizer and then we collected as many bags as we could of plant parts that we knew might be active like leaves, bark, flowers, fruit, et cetera.[56]

Elaine Hoagland, recent president of the Association of Systematics Collections, is particularly worried about such practices, as they threaten the entire future of systematic collecting. Source countries have suggested that they may prohibit all forms of biological collecting, either scientific or commercial, if they feel they are unable to make a meaningful distinction between the two. In 1995, following reports that material collected on a scientific expedition in the Pacific had since been accessed and patented by a commercial pharmaceutical company, a number of Pacific island countries threatened to impose an embargo on all collecting in the region and to declare the Pacific "a lifeforms patent-free zone."[57] Commenting on this situation in 1996, Hoagland expressed considerable concern: "This is so dangerous to the systematics community because it could potentially cut out all collecting because they can't distinguish between biodiversity prospecting and the poor graduate student who wants a degree in some exotic group of beetles. So this is a real problem for us."[58]

It could be argued that scientific institutions have bought this situation upon themselves by engaging in a practice that inevitably blurs the boundaries between academic, scientific, and commercial activities. This research suggests, however, that the decision to allow commercial interests to access publicly funded scientific or academic collections is driven in large part by external changes in the broader U.S. economy. Progressive cuts in public funding, increasing corporatization, downsizing, and the subcontraction of specialized services are all beginning to have their effect even in the apparently rarefied atmosphere of universities, botanical gardens, and museums of natural history. Publicly funded institutions are finding that grants for basic "scientific" or "academic" research—including biological-collection programs—are being slashed, placing them under an intolerable financial pressure to secure income from alternative sources. In order to continue with research that they consider to be essential to the broader project of biological conservation—species identification, taxonomy, and surveys of rates and patterns of extinction, for example—they have resorted to selling access to their collections or contracting out their services as commercial collecting agents.

Since the mid-1980s, many such institutions have agreed to enter into contractual agreements with companies in order to fund basic research or collecting programs. The Smithsonian, for example, entered a contractual agreement with the company Pharmacognetics in 1996. Under the terms of this agreement, Pharmacognetics agreed to underwrite part of the cost of the botany department's scientific collecting program in exchange for access to material collected on these expeditions. The curator of botany responsible for negotiating the details of the contract has clearly been exercised by contemplation of the conflicts of interest that arise when publicly funded "scientific" researchers are seen to spend their time collecting for private corporations. Ironically, in explaining what this agreement is not, he actually gives a very accurate description of exactly what it is: "We just started this and it took about a year to draw up the contracts because we are a federal group and we didn't want to look like the taxpayers' money was going to support our salaries and that we were then renting ourselves out—or our services out—to a private company because that's not what it is. Our research funds are dwindling because federal funds are being cut."[59]

Although the establishment of this formal subcontractual relationship with Pharmacognetics might alarm some of the benefactors of the Smithsonian, there is again nothing illegal, or indeed even unethical about it, provided of course that it is subject to established protocols and public scrutiny. Pharmacognetics initially made a public commitment to share any profits that arose from the use of the material with supplying countries. However, there is, of course, no certainty that the "suppliers" of such materials are either aware that material collected by the Smithsonian (an eminent, internationally respected scientific institution) could be circulated to Pharmacognetics for commercial use or that they would give their consent to such practices if they were informed. Furthermore, in the absence of a direct contractual agreement between themselves and Pharmacognetics, they have no legally recognized right to trace what Pharmacognetics does with the material or to whom they might circulate it or to recover royalties for its use. Pharmacognetics has since decided to defer using samples collected through the Smithsonian until access agreements conform precisely to requirements under the biodiversity convention. In the interim the company will only collect in the United States and dependent territories.

The establishment of such close relationships may also invite the commercial partner to attempt to access other material in an institution's broader collection. The curator of the Smithsonian's botanical collecting program recently confirmed that such a scenario was entirely possible, that Pharmacognetics may, at some point in the future, show an interest in bioprospecting other sections of the Smithsonian's existing collection, or even those held by other publicly funded institutions to which the Smithsonian has links:

> Right now the samples that we provide for them are just from our research collections, but I could envision a situation where we could interact with other institutions, like the NYBG or Missouri, but right now we are only working with field-collected samples from the host country. As I said earlier, we have a greenhouse full of material that our researchers have collected from various parts of the world over the last ten years before this agreement was established. We could also sample those things [for Pharmacognetics or another company], but that is a little trickier now as those things were collected before the treaty, and that has some ramifications that we don't want to get into. So at this point it's all still focused on the material collected on field expeditions, but that could change.[60]

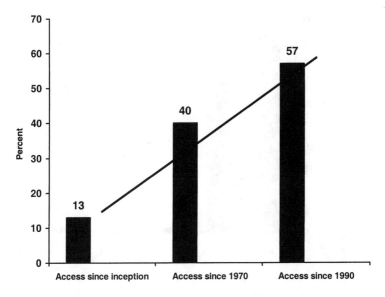

5.1 Percentage of respondent institutions that allow commercial access to their collections (n = 45).

THE EMERGING TRADE IN COLLECTED GENETIC AND
BIOCHEMICAL MATERIALS

Although the practice of allowing commercial interests to access existing scientific or academic collections in exchange for fees or funding appears to be increasing, there has as yet been little empirical research to confirm this. In order to assess how widespread these practices are, I obtained permission from the Association of Systematics Collections to survey their membership. The association comprises eighty-three academic or scientific institutions, the great majority of which are in possession of systematics collections.[61] The findings of this survey confirm that the practice of allowing commercial interests to access these collections in return for payment or some other form of funding or infrastructural support is increasing rapidly. Of the eighty-three institutions surveyed, forty-five returned a long questionnaire that provided detailed information about their past and present practices. Of these respondents, 53 percent reported that they had noted a marked increase in commercial interest in their collections. This interest was reported to have come from a variety of quarters. Many companies had approached institutions requesting specimens, stating that they needed them for identification purposes or in order to establish their geographical distribution.

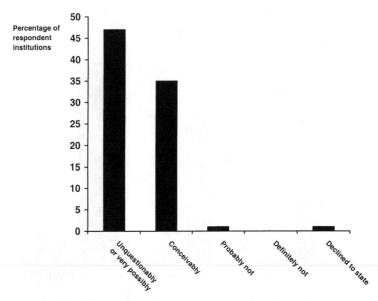

5.2 Respondents' assessment of whether institutional collections will become more commercially valuable.

This may be have been for bioprospecting purposes; however, it may also have been for ecological, archaeological, or even forensic research. A small number of respondents stated that they allowed companies access to their collections in order to photograph, draw, or film the specimens within it.

Given the sensitivity of institutions' allowing commercial interests to access their samples of material for the purposes of genetic investigation, tissue sampling, or cell culturing or screening, it is perhaps surprising that 20 percent of all respondents stated explicitly that they had allowed these specific activities to occur. An even larger proportion of respondents stated that they were allowing commercial access to the collection but declined to state for what purpose. In all, 66 percent of all the respondent institutions stated that they allow commercial access to their collections, 13 percent of which have done so since inception, 27 percent of which have done so since 1970, and a further 17 percent of which have done so since 1990 (see figure 5.1).[62]

There is no question that curators are well aware of the changing uses to which such collections may be put. Indeed, 42 percent of all respondents explicitly stated that they considered collections to be a valuable source of archived or extracted DNA, cultivars, or tissue samples. A breakdown of the figures relating to the potential or future value of the collected materials re-

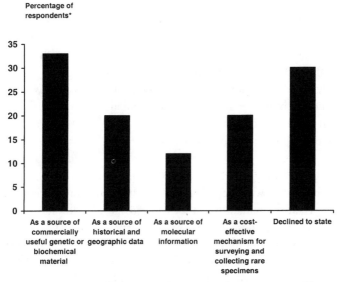

Percentage of
respondents*

*Stating that the collections would unquestionably or conceivably become more commercially
valuable (No=39). Percentages do not total 100 because some respondents gave more than
one reason

5.3 Reasons why respondents believe institutional collections will become more commercially valuable.

veals that 47 percent of all respondents believe that their collections will either "unquestionably," or "very possibly," become more commercially valuable in the near future. A further 35 percent believe that they could "conceivably" become more commercially valuable. What are significant however, are the reasons that they gave for arriving at this conclusion. Their written responses clearly indicate their belief that the collected material is already, or will become, an important source of raw material for the pharmaceutical or agribusiness sectors. Of the 82 percent of all respondents who stated that the materials would in all probability become more commercially valuable, over half of them (57 percent) made explicit references to this particular use of the material (see figures 5.2 and 5.3).

In naming the sections that they believe will prove most valuable in the short or long term and in explicating the reasons why, respondents wrote, for example: "the cell cultures—for their clinical applications and use in biological production"; or "the living collection—they can provide tissues for cultivation or testing"; or "the insect and plant collections—we in the scientific collection community are already seeing increased interest in collections as data and material sources for medical and agricultural biotechnology." Many also made reference to the advances in technologies that were facilitating this process, for example: "New technology will enable DNA to be extracted and amplified to degrees useful for pharmacological analysis from the very large components of dried and liquid-preserved botanical specimens"; or "DNA can already be extracted from dried plant collections as old as 90 years in age, so as technology improves there is greater potential for getting at the genetic material in the collections."

The written responses also indicate that these collectors are conscious of the range of reasons why commercial interests would engage in microsourcing, including the difficulty and costs associated with in situ collecting. When asked why they thought the collections would become more valuable, several responded that "they can provide samples that are rare or that are from sources whose distance makes it expensive to get"; "sample collection and identification is extremely expensive and technological advances will allow us to utilize preserved specimens for many types of testing and study"; and "genes will be dollars and companies will look to find them as cheaply as possible." Of the 98 percent of all respondent institutions who circulate material from their collections to other similar institutions, researchers, or corporate concerns, the vast majority (92 percent) admit that they have, at present, no way of monitoring how that material is subsequently used. The majority of samples or specimens are lent to other museums for the purpose of taxonomic identification and other similar tasks, but others are sold to commercial interests, with one respondent confirming, for example, that they will sell samples: "charg[ing] 50 dollars to commercial users or to researchers who cannot exchange tissues."

Despite the increase in commercial access to and use of their collections, those working in the institutions involved seem reluctant to fully acknowledge the implications of these practices. Many display an understandable unwillingness or inability to recognize that they are now *also* engaged in a commercial enterprise and a new trade. For example, one institution responded that they did not allow commercial interests to access material in the collection, but in response to another question asking them if they received compensation for providing material from the collection to outside users, they responded, "No—unless it's for commercial purposes and then we negotiate a price for data or samples." Another responded that none of their collection was for commercial use "except for that collected for bioprospecting." Others respondents have denied that there has been a recent increase in commercial interest in their collection while confirming else-

Table 5.1   Nature of the increasing commercial interest in institutional collections of biological material

Nature of the request or offer made by commercial interests	Number of requests or offers made to surveyed institutions	Number of requests or offers accepted by those institutions
Requests for samples or specimens from the collection	15	7
Requests to examine voucher specimens held within the collection	11	6
Requests to access information about the collection held on institutional databases	17	10
Offers to provide commercial contracts to collect	8	8
Offers to fund existing collection programs in exchange for the collection of specific material	3	1
Offers to fund existing collection programs in exchange for access to the wider collection	4	1
Offers to fund an internship or placement within the institution	5	1
Totals	63	34

where in the survey that they have not only been made offers to collect for companies but that they have, in many instances, accepted them. In fact, 45 percent of all respondents were able to list specific offers that had been made by different companies with interests in their collection that had since been accepted by the institution. These twenty institutions have, between them, received and/or accepted over sixty recent offers of interest from commercial companies (see table 5.1). It is precisely these practices that blur the distinction between scientific, academic, and corporate entities and activities and that inevitably undermine the scientific status that these institutions so jealously guard.

Crucially, however, the survey confirms that the institutions that have taken up such offers have done so as a direct consequence of their need to secure alternative sources of funding. Indeed, 31 percent of respondents specifically noted that this was one of the advantages of allowing increased commercial involvement in, or access to, their scientific collections. They remarked, for example, that it would provide "increased funding in an environment where many collecting-based institutions are in difficulties." However, 25 percent of all respondents recorded serious concerns about the effect that accepting such offers might have on their status as scientific research institutions or the direction of their research agendas, stating, for example, that "collections will come to depend on studies of commercial species and taxonomic studies may suffer as a consequence" or that "the disadvantages are the commercial determination of collecting priorities rather than scientific decisions." Curators are also aware that the commercialization of the material may result in patent claims that could inhibit basic research. One responded, for example, that a major disadvantage of allowing corporate access was that "some material may no longer be available to the scientific community due to ownership or information release restrictions."

Scientific and academic collecting institutes are not the only organizations that are engaging in such practices. Conservation agencies are also surviving in a harsh economic climate by becoming increasingly corporatized. Organizations such as Conservation International, among others, have received public funding to undertake research into ethnobotanical uses of biological materials on the understanding that it will benefit indigenous communities in developing countries. It was on this basis that CI received funding to facilitate the collection of specimens and information in Surinam under one of the ICBG agreements. At the same time, however, conservation agencies are increasingly turning to corporate enterprises to fund their operations. In 1995, CI entered into a commercial partnership with the cosmetics corporation Croda International. Under the terms of this agreement, CI was paid to facilitate the collection of ethnobotanical information and biological specimens on behalf of Croda. Croda, in turn, intends to use the collected samples and information in the manufacture of personal-care products. CI has argued that their "entrepreneurial approach," which "applies business models to conservation," creates new opportunities to develop biological resources in an equitable and sustainable fashion. While this may be correct, it does so by creating new links and partnerships between formally disparate actors such as conservation agencies and corporations, creating new conduits though which samples may be circulated, exchanged, or traded.

The accelerated circulation of these materials need not necessarily be a cause for concern—as long as there is some means to ensure that all subsequent uses of the material can be effectively monitored and will benefit the source country in some way. Although every public assurance is given by agencies such as CI that they have the capacity to trace the successive uses that are made of collected materials, an analysis of recent practices suggests that commitments are not always fulfilled. Bio-informational proxies are fragmentary in nature—tissue samples, cell lines, and libraries of extracts are easily transported, exchanged, copied, even lost. They are, by nature inherently difficult to trace. Moreover, as engineered artifacts, they always remain subject to claims of private ownership. The (sometimes specialized) labor invested in their manufacture is such that the "collector" may in fact regard himself as an "inventor" with an inalienable right to possess the collection that he sees as an embodiment of his intellectual labor. In an interview, I asked Lisa Famolare, a program director at Conservation International, about the fate of an earlier collection of samples and indigenous information about their use that had been generated by an ethnobotanist working for them, under a program that CI had instituted in Costa Rica in 1993. As she notes, so clearly, the collection has been copied, relocated, almost certainly reused, or at least retained for future use, although how and in what ways cannot be established with any certainty even by those whose direct responsible it is to know. I began by asking her what had happened to the collection:

> **L.F.:** Good question. I know that the people [in Costa Rica] have a copy of it, but I'm sure that X has a copy of it, but I don't know what he's done with that.
> **B.P.:** But it is held by him and not by CI?
> **L.F.:** Yeah, he took that, since he doesn't work here any more. He has that collection now, and he has another foundation that he's developing.
> **B.P.:** Is that a conservation foundation, too?
> **L.F.:** Yes.
> **B.P.:** And how have the benefit-sharing agreements played out over time?
> **L.F.:** I really don't know—you'd have to talk to him about that.[63]

Executives working in the pharmaceutical industry have expressed some skepticism about the role that conservation agencies play in bioprospecting initiatives, drawing attention to the way in which they can employ their nonprofit, charitable status to secure public or private funding that enables them to reduce the costs associated with in situ collecting while amassing a valuable collection that they are free to exploit commercially in whatever ways they see fit. Commenting on such practices, King remarked:

> I don't know how much money they [conservation-agency partners] get directly [through federally funded programs]—I mean I think

that's an important question—it's a great question. I would really like to know what kind of [biological] materials they're getting compared to what cost. They may well be getting the deal of the century. Some of the criticisms are that they are using government funding and public money to collect material that they then sell as a commodity. On the other hand, you can't link development and conservation unless you've got a link—but, you know . . . [64]

HIRE PLANTS: RENTERS AND BROKERS

As the biotechnological revolution has advanced, molecular approaches to the development of pharmaceutical and other therapies have become more commonplace. Consumers are interested in securing samples of material, in part, so that they may derive from them genetic or biochemical material or information about the structure or action of these extracted elements. This information may be rendered in a number of different ways—it may remain corporealized, existing as an extracted sample of DNA, for example, or it may be entirely decorporealized, existing only as sequence information stored on a database. As I noted earlier, information is an unusual commodity in that it can be given away and yet still possessed. This phenomenon has resulted in the creation of particular modes of transaction. Those in possession of information or information-based products have recognized that they may either sell their products to consumers outright or, alternatively, sell access to the information embodied within their products *repeatedly* on a rental or "pay-per-view" basis. This has proven to be a particularly lucrative means of capitalizing on patented or copyrighted information, as the owners of many of the world's largest media conglomerates will attest. People who wish to use the information once only or on a "try before you buy" basis are creating a large consumer base for these services.

Given that a large number of consumers are now also interested in securing access to a variety of different types of bio-information—sequence information or information about molecular or biochemical structures, organization, or function—it should come as no surprise to discover that new forms of commodity exchange are evolving that mirror those that characterize the exchange of other types of information and information-based products. A number of companies that have emerged in recent years—they might be thought of as brokers—specialize in creating libraries of plant, animal, fungal, microbial, and even human genetic materials, extracts of chemical compounds from plants, marine animals, and microorganisms, and databases and DNA-sequence information to which they intend to sell access on a rental or "pay-per-view" basis. The notion of a "rental firm" perhaps conjures up an image of a small commercial enterprise akin to a local video store. However, the kind of organizations that intend to specialize in the rental of such material are surprisingly diverse, ranging from small commercial enterprises, such as the Knowledge Recovery Foundation, to the corporate subsidiaries of publicly funded universities, such as the Strathclyde Drug Research Institute. Each operates on the premise that collected libraries of genetic or biochemical material can be utilized and reutilized by a number of interested parties. With each transaction, the material resource is screened for different types of biochemical or genetic information embodied within it. Each recipient of the material may be looking for different types of information and, as a consequence, the same material may be recirculated a number of times before being wholly consumed.[65]

Some of the largest collections of material that have been created over the last decade are increasingly being used in this way, even if those in charge of them are still slow to recognize this fact. The National Cancer Institute, for example, has recently made a decision to allow outside interests to access the material stored in their repository. The NCI will not profit directly from this scheme; they do not charge these companies for the right to test the material, although they could. Their intention at present is simply to maximize opportunities to find companies interested in the commercial development of the materials. The materials are transferred to a range of third parties under a material transfer agreement. If a company finds a promising lead, the NCI then directs them to the original supplier of the material, leaving it up to these two parties to arrive at a mutually satisfactory agreement on appropriate compensation. The NCI plays no role in these negotiations. In an interview, Gordon Cragg described how the scheme works:

> We have this program because we have got this huge repository of extracts and the general feeling is that this is a fantastic resource . . . a very valuable sort of national resource . . . which other organizations should be able to screen for activity, but we do this under an agreement which is a legally binding document. One of the main clauses in that agreement is that should anything of interest emerge from that then they [the recipient company] must follow all the terms in our Letter of Collection relating to the source country

where that organism originally came from, and if a great drug comes out of it they've got to negotiate with that country the terms of compensation. We've got a list of the organizations that do have access—there are seventeen or so now, I think. Of course Merck is the biggest, but there are also a number of smaller companies which specialize in high-throughput screening, and they aren't necessarily pharmaceutical companies—they just do a lot of screening, and they could either try to work them up themselves or, if they get interesting leads, then develop interactions with other industries. Then there are the universities—University of Mississippi, for example—they have a natural-products research center and they have access to our repository. . . . That is what is happening at the repository. We are being encouraged by our new director to be, how shall we say, more interactive with the extramural community, to really encourage access because this is regarded as a very valuable resource which should be utilized to the maximum, but of course, as I say, I want, while we are doing this, to make sure that we protect the rights of all of the countries where all these organisms come from.[66]

Collections such as that stored in the NCI's repository are, as Cragg notes, extremely valuable, in part because they are costly and time consuming to assemble. Moreover, as the evidence presented here suggests, it is extremely unlikely that a resource of this scope and size could have been generated without government funding. This raises the question of why commercial users should be able to access this resource without paying a fee. All users must pay a handling charge, but they pay nothing for the right to test, screen, or examine the material. Such practices are not restricted to the NCI repository. Those that have been involved in creating other collections using public monies are also considering how it might be possible to capitalize on these resources both now and in the future. Joshua Rosenthal, the director of the ICBG program, has confirmed that several of their consortia are looking to devise new mechanisms to reutilize collected materials once their initial exclusivity agreements with their existing commercial partners expire:

> Each of the consortia has anywhere from three to nine cooperating institutions, and each has made different decisions about what will be done with the extracts that are made and the bulk collections in the long run. The question has come up for the groups that are fur-

thest along, How can we make use of the 99 percent of collections that have been found to be not of commercial interest to the commercial partner and/or the university researchers—because they are negative for the diseases they are looking at or duplicative of things they already know about? And what one of the groups has done, and what I think a lot of them will do, is to try and collect enough material the first time out—like a kilo or two of material—so that after the period of exclusivity expires, they could develop a bank of residual extract materials and residual bulk collections that can be remarketed in other agreements with other researchers. They haven't figured out quite how they are going to do that, but the pharmaceutical partner, BMS in this case, has offered to help them market them elsewhere.[67]

Other, privately run organizations have been set up with the sole intention of collating materials or information that can be offered to consumers on a strictly commercial "pay-per-view" basis. The Knowledge Recovery Foundation, a New York–based company, undertook its own collection programs in China and Peru with the aim of creating focused libraries of tissues and extracts that can then be rented out to interested companies for testing or examination. The company offers the consumers only limited taxonomic and geographic information about the samples, operating on the assumption that the consumers, having "paid to view" the material will then have to return to the foundation to acquire the information necessary to obtain further supplies of the material should they require it. It is at this point that an agreement on price will be negotiated, although there is no guarantee, of course, that the price will include any fee or compensation for the source country. Other companies are intending to act as brokerages or clearinghouses for collections of "rentable" materials. Companies such as Pharmacognetics, for example, seek to acquire collections of tissues or extracts (though institutional, corporate, or organizational contacts) that they can then license out to other interested third parties. John Kress, who works at the Smithsonian and has collected plant material on behalf of Pharmacognetics, described the principle on which the company operates:

> Pharmacognetics came to us and said, "We are business people trying to start a bioprospecting firm, and we need some botanical expertise to help us make the collections which we will use for commercial purposes." We [the Smithsonian] are part of the government, and we have to be somewhat careful about how we make use of our

resources, and so the agreement that we came up with is that this company Pharmacognetics, which is a start-up company—they're not a Merck or any of those—will essentially pay for our fieldwork research in return for botanical tissue samples that we collect for them. They want to serve as a sort of clearinghouse for botanical tissue—for any other company that is looking for botanical tissues to use in other areas such as agriculture, pharmaceuticals, cosmetics, et cetera. So they are not doing a lot of the screening themselves; they are just bulking up the tissues and maybe doing certain extractions but that's where it stops.[68]

Companies that are interested in acquiring a "rentable" collection of material may look to secure it through means other than in situ collection. They may, for example, approach a brokerage firm or an individual broker that specializes in the buying and selling of whole collections of material as well as the sale of individual specimens or extracts. Restructuring within the pharmaceutical industry has led some corporations to move out of natural-products research, and their collections of extracts and specimens have been sold on the market as assets. University laboratories are also rationalizing their research priorities and selling off collections of extracts in order to raise revenue. In each case, brokers have been engaged to facilitate the sale and resale of such material. During the course of my research, I interviewed one such broker who works, on a freelance basis, buying and selling these "orphaned" collections of genetic and biochemical material. She explained her role in a recent process of exchange:

I was approached by this New Zealand university that had a library of biochemical extracts that had been developed under an earlier research program by a department that was being closed down. They thought the collection was probably useful, so they asked me to try and find a buyer for it. They had four copies of the extracts, and so I'm trying to sell them on an exclusivity basis where the first buyer gets exclusivity for a year, then I can sell the second copy to another company and so on. It looks like I've already found a buyer for the first copy, so that's good. They told me that they didn't think any of the material was originally collected on . . . Maori land or anything, so that shouldn't be a problem.[69]

As Neil Belson, the president of Pharmacognetics, so eloquently noted, it is not necessary for brokers to sell these samples outright to other interested third parties. That would result in a loss of the asset. They have realized that they may capitalize on these collected resources much more effectively by allowing a series of interested parties to view them successively, while charging each one for the privilege:

> You don't have to sell the samples—you could basically license them out for experimental purposes only . . . rent it to them, and then if they find something interesting, they can work with you on doing something more with that sample. . . . The agreement would say that for x dollars you are getting the right to test this material and to have first right to negotiate further collaboration with us if you find something of interest. That's how it would work.[70]

In the speculative world of biotechnology, many consumers are prepared to risk paying to examine material or information that shows potential. If only one set of material or information goes on to form the basis of a highly profitable product, the returns for the sale of that product would more than compensate for the cost of accessing other sets of material or information that may ultimately prove to be of little utility.

TRANSACTING BIO-INFORMATION: LICENSING AND "PAY-PER-VIEW"

It is not only samples of plant or animal material that are being accessed in this way. As genome sequencing advanced in the 1990s, it became possible to generate new comprehensive databases of interactive genomic information. This information, which had previously remained embedded in particular organisms (bacteria, viruses, plants, animals, and humans), could suddenly be rendered in a new digitized form. When translated into data, the information could be viewed and accessed with even greater ease—and, more importantly, circulated instantaneously from one location to another no matter how great the distance. These developments inevitably resulted in the creation of a new resource economy or trade in digitized bio-information. Once again, consumers pay to gain access to this information through mechanisms such as licensing and pay-per-view. Bill Neirmann, former research director at the American Type Culture Collection, explained how this nascent trade in genetic sequence information operates:

> We work locally here with Human Genome Sciences and The Institute for Genomic Research. One is a for-profit, one is a nonprof-

it, and the whole thing was set up by Smith Kline Beecham pharmaceutical company. They [SKB] are basically looking at establishing sequences of human genes, and they're setting up this huge
database of these kinds of gene sequences. And so what they are doing is licensing people to search it for things—to pay money to
look. And as part of the agreement, you can look, you can find
things, you can commercialize it, but Smith Kline gets first right of
negotiation with you on commercial arrangements. So the details
of the profit sharing are not established up front, just that you agree
to negotiate first with Smith Kline.[71]

The amount of money that may be generated through pay-per-view access to genetic databases is substantial. In February 2000, Incyte announced
that it had signed access agreements with Pfizer, Eli Lilly, AstraZeneca, and
other major drug manufacturers, giving these firms access to its database of
genetic information and data-management software tools. More significantly, the company announced its intention to fully exploit the informational
aspects of their products by creating a new internet strategy, designed to enable researchers (both corporate and academic) to tap into the database online from their home institutions, wherever they might be. In announcing
the decision, the company's director revealed that the total life-science research budget of U.S. academic institutions alone amounts to $18 billion,
about twice that of the global pharmaceutical industry. He noted that the
company expects to capture a sizeable portion of this budget in return for
access to their databases. In the wake of this announcement, Wall Street an-
alysts raised their twelve-month price-target estimates for Incyte from $235
to $374 per share, an increase in value of some 60 percent.[72]

These developments raise some important questions. A principle was established under the biodiversity convention that enshrines the right of
source countries to receive some form of compensation for the genetic and
biochemical materials that they provide for the purposes of research and development, on the understanding that these resources form part of the patrimony of each nation-state. At first remove, when the material remains in a
largely corporeal form, this principle is generally adhered to and the payment of some sort of compensation is usually agreed. However, *as the material begins to travel and unravel,* with each successive decorporealization,
each successive use, each successive remove, this commitment weakens a
little further. The less material, and more informational, these resources become, the poorer, it seems, is the probability that source countries will re-

ceive any compensation or return for their use. This is a curious phenomenon, given that these bio-informational resources are clearly highly valued and highly sought after. As the evidence that I have presented here suggests, companies and brokerages secure a financial return each time these bio-informational resources (in whatever form) are transacted; however, it is not clear that they have any effective means of tracking these transactions or of ensuring that some form of compensation is returned to the supplier for each successive rental of the resource.

Even in instances where brokerages intend to ensure a return, they may find themselves dogged by a risk that attends all information-based rental transactions: unlicensed copying. As Neil Belson suggested in an interview, there is always a concern that the recipient will simply copy, clone or synthesize, the resources: "In the best-case scenario, we would market tiny amounts of material to pharmaceutical companies, enough for them to tell whether the material warranted additional investigation for their study but not enough that they could take it and run with it and never need us again."[73] Given that the amount of material required for processes of elucidation and replication is reducing all the time, it is difficult to imagine how brokers can be certain that unlicensed copying is not occurring. Preventing the unlicensed copying of other types of information-based products—such as computer programs or samples of music—has proven to be an extremely difficult task, as the review in chapter 3 revealed. This raises the question of whether there is, in fact, any effective way of monitoring how collections of genetic or biochemical materials are used by successive recipients. If such material were to form the basis of some new compound or used as a starter for processes of replication or synthesis, would the broker, let alone the source country, ever know? Is there any reason to assume that source countries will receive any compensation for the use of material circulated or exchanged in this way?

It has been argued that speeding up the rate of exchange of these informational resources can only advantage supplying countries, as they will benefit directly from this trade—by either being invited to supply further samples of material or through the direct receipt of monetary compensation. As I have illustrated here, the probability of source countries being asked to provide ongoing supplies of raw material is diminishing rapidly. Advanced technologies are unevenly distributed—while many developing countries lack the infrastructure necessary to capitalize on their own bio-informational resources, collectors are able to transport these resources to centers where they may be utilized to great effect. The fact that the material and information that they seek is now available to them in new forms—as cell lines, tissue samples, DNA sequences, and the like—only serves to make these valuable resources easier to collect, concentrate, control (as alienable commodities), and recirculate. Suppliers' only hope of securing compensation rests on the probability of their receiving a portion of the income that companies derive from their successive use of these bio-informational proxies. For this to be the case, it would be necessary for the protocols that govern the exchange of such material to be adequate to the task of tracking multiple transactions involving a resource that is inherently fluid, mutable, and dynamic. The question of whether such protocols can ever be truly effective in this regard is addressed in the final chapter.

In this chapter, I have investigated the changing ways in which collections of biological materials are used, traded, and exchanged as industrial commodities within the American pharmaceutical industry. Although it has for some time been possible for companies to reproduce biological materials using processes of artificial replication, technology has rarely been sufficiently advanced to make this a viable means of production. Evidence produced here suggests that this situation is now changing. Advances in techniques such as plant-cell culture have made it possible to reproduce viable quantities of genetic and biochemical materials from milligrams of collected material. As a consequence, collected samples of material can now be used as a means of production for that material even if they do not have "naturally" regenerative capacities.[74] Moreover, the development of new techniques such as combinatorial chemistry now enable collectors to recombine collected genetic and biochemical resources in order to form new patentable products. This provides some important supporting evidence for Castells' argument that biotechnology ought to be characterized as an information technology on the basis of its ability to "decode and reprogram the information embodied in living organisms." Evidence also suggests that while technological advances are reducing the need to collect large quantities of material, the requirements for benefit sharing established under the biodiversity convention are, in some instances, also creating disincentives to collect.

These developments will have a considerable impact on the social and spatial dynamics of collecting, notably on what is collected and when. While it will still be necessary for companies to collect small quantities of material that could provide necessary "lead compounds," large-scale, in situ collecting is unlikely to continue. Evidence from this research suggests that some companies are finding it more expedient to "microsource" samples of genetic and biochemical materials from existing collections. This inevitably has the effect of creating a new market economy in bio-informational resources and new opportunities for trade in the same. The new linkages that have been created between organizations described in the previous chapter provide important new avenues of exchange for these materials. This study provides clear evidence that scientific and academic collections in the United States are being "remined" for genetic and biochemical resources. It also provides evidence of the new forms of commodity exchange—such as rental, licensing, and pay-per-view—that allow collectors to capitalize on the recirculation and reuse of collected samples of material. This raises the question of who will benefit from these successive uses. Will the regulatory controls and compensatory agreements established under the biodiversity convention prove adequate to the task of tracking the uses that are made of these bio-informational resources over both space and time? If they are not, what implications will this have for the rationales for bioprospecting established under the convention: namely that bioprospecting will provide an ongoing source of income for suppliers in developing countries and hence an important incentive for preserving endangered environments? These questions come under investigation in the following chapter.

6. TAMING THE SLIPPERY BEAST
REGULATING TRADE IN BIO-INFORMATION

As a consequence of the introduction of new technologies such as photo-copiers, digital scanners, computers, satellite communications, cable televi-sion, and the Internet, it has become possible to render the information that has historically been embodied in particular artifactual forms—as books, paintings, documents, and the like—in new artifactual forms—as software files, digitized images, and electronic data. Liberated from the physical housing in which it was previously embodied, this information becomes footloose—able to be circulated, copied, archived, and recombined at speed and with comparative ease. Fluid and mutable, it is free to travel along the many capillaries that constitute the increasingly dense filigree of connec-tions that make up the new global information network. However, these very factors have also served to make the task of monitoring and regulating the millions of transactions that occur within these new "spaces of transmission" a particularly difficult one.

The introduction of new biotechnologies has similarly allowed biologi-cal materials to be rendered in new ways—as cryogenically stored samples of tissue, as active biochemical extracts, as cell lines, or even as sequence information. These new proxies privilege the informational content of the biological material at the expense of much of its corporeality, which is sub-sequently divested. Much more lightweight and mobile than the whole or-ganisms from which they are drawn, these proxies may also be circulated, copied, archived, and recombined at speed and with comparative ease. As the evidence presented in the previous chapters suggests, several factors— a progressive move towards the molecularization of approaches to the de-velopment of pharmaceutical and gene therapies, the decision to construct genetic and biochemical resources as alienable commodities, and the cre-ation of new interlinkages between consumer groups—have all combined to create a flourishing market for these bio-informational proxies and a multitude of new conduits through which they might be traded or ex-changed. Genetic and biochemical information, it could be argued, has be-come considerably more accessible, transmissible, and replicable than it was when embedded in whole organisms.

It has also, as a direct consequence of this, become much more elusive. Able to be cloned, copied, synthesized and engineered, rented, downloaded, viewed, and exchanged, these bio-informational proxies may be transacted lit-erally thousands of times in any given month or year. If we accept the princi-ple that in this technoscientific age, what may be of most value and requisite of most protection in a biological "work" is the genetic or biochemical "infor-mation" embodied in that work, then it stands to reason that the protocols that are put in place to prevent unlicensed, and therefore uncompensated use of the work must extend to cover all the successive uses that are made of that ma-terial or information, in whatever form it is rendered. How, though, will it be possible to keep track of all the uses that are made of collected genetic and biochemical materials and information? It could be argued that just as processes of "dematerialization" have undermined attempts to keep track of the successive uses that are made of copyrighted information, so processes of "decorporealization" also have the capacity to derail attempts to regulate the successive uses that are being made of bio-informational resources.

The Convention on Biological Diversity established the principle that the suppliers of genetic and biochemical "resources" should receive a por-tion of the benefits that accrue to others through their collection and uti-lization. However, the convention's text provides only a broad interpretation of what constitutes "genetic resources," defining them only as "genetic ma-terial of actual or potential value." It is not even immediately evident how the term "material" might be construed in this context—would it include only those resources that retain some degree of corporeality, or might it in-clude sequence information, for example? The use of the term "of actual or potential value" suggests that derived genetic and biochemical materials and information should be construed as recompensable resources, even in cir-cumstances were their utilitarian value has yet to be established. The com-plexities of determining for what compensation is due, how levels of com-pensation might be set, or to whom it should be paid is further complicated by the fact that the convention provides no explicit indication of how the principle of benefit sharing might be translated into workable contractual and/or compensatory agreements.

Over forty countries now host bioprospecting operations, and each is at liberty to devise its own contracts and compensatory agreements. Despite this, few of them have introduced compensatory agreements that deviate in any significant way from the three-phase compensatory model first devised and introduced by the NCI in 1988. Although quite a deal has been written about compensatory agreements over the last decade or so, much of this has been authored by those collectors, organizations, or institutions that have been directly involved in generating them. The reports consequently tend to be more descriptive than analytical. They provide information about the terms and conditions of the agreements but offer little insight into the role that different actors played in their creation or of the power relations that exist between the different parties. A curious opacity thus still surrounds the history of the construction of such agreements. Little is known about who lobbied for their introduction, who authored them, whose interests they apparently and actually serve, and why particular agreements have become hegemonic over space and time. In the interests of remedying this situation, I devote the first section of this chapter to a detailed investigation of these issues, employing evidence from my empirical research to disclose the nature of the machinations that have underpinned the development, implementation, and proliferation of these compensatory agreements.

Having contextualized these compensatory agreements within the milieu that produced them, I then consider the question of how effective they are likely to be. In undertaking this assessment, I examine the basic three-phase compensatory model that has been adopted so widely, concentrating specifically on the central but much neglected question of how well such agreements will cope with the spatial and temporal challenges that are inherent in the governance of all informational resources, including bioinformational resources. Tracing the journeys that these proxies make as they peregrinate through multiple reworkings and along many avenues of exchange will undoubtedly prove to be a very challenging task, as regulators in the computer software, global finance, recorded music, and other information-based industries have discovered. They have dedicated many billions of dollars to attempting to track and regulate the flow of highly replicable and transmissible information-based products (such as DVDs and MP3 recordings) over the past decade with questionable success. Bioinformational resources are now subjected to an equally complex range of transactions, and it is unlikely that these will prove any easier to monitor or control. By drawing attention to these similarities, I hope to raise a series of questions about the wisdom and indeed the viability of investing serious amounts of time, energy, and finance into what I see as an inherently flawed activity—attempting to trace the successive uses that are being made of biological derivatives, including types of bio-information. Given that compensatory agreements rely in large part on the successful prosecution of this project, any failure to achieve this goal must threaten their viability as a mechanism capable of delivering the kind of distributive justice that their authors hoped to achieve. My ultimate aim in revealing the inadequacies and limitations of these agreements is to direct our collective attention forward to a consideration of the question of what compensatory paradigms (if any) might be employed to adequately remunerate suppliers for the collection and use of this elusive, slippery beast—bio-information.

COMPENSATORY AGREEMENTS:
THE RISE OF A PROTO-UNIVERSAL CULTURE OF REGULATION?

The first attempts to devise a formal compensatory or benefit-sharing agreement relating to the collection and use of nonhuman genetic and biochemical materials were made in 1987 and 1988 by those working in the Natural Products Division of America's National Cancer Institute. In relaunching their collection program in 1987, the NCI established a new, and extensive, network of collectors by enrolling a host of publicly funded research institutes, botanical gardens, and universities in both developed and developing countries into their bioprospecting operation. These collaborating institutions undertook the collections in the field, providing a first, and ongoing point of contact for targeted communities in source countries. In 1987, word began to filter back to the NCI's headquarters, via these collaborating institutions, that supplying communities were dissatisfied with the NCI's collecting activities. Their primary concern was that they were not receiving any compensation or financial return for their collected materials and that these projects were, in principle and in practice, indistinguishable from those undertaken during an earlier, and more exploitative colonial era. Having been sensitized to this argument, the collaborating institutions began to lobby the NCI to develop a formal compensatory agreement for use in source countries. As Gordon Cragg explained in an interview, it was researchers working at the New York Botanical Garden who first encouraged the NCI to devise its "Letter of Intent.": "This policy, we didn't have it to start with in 1986, but due to the sort of advice of folk like Mike Balick and the other collectors who said these countries were justifiably concerned about what would happen to their material, we decided we better do something about it, and so the "Letter of Collection" evolved from a fairly small beginning to something pretty major."[1]

Prior to the development of the NCI agreement, there were few controls on the collection or exportation of biological specimens or samples, excepting those that were classified as exotic or endangered. Companies or organizations that wished to acquire specimens would either send their own scientists on field expeditions or, alternatively, contract operatives in source countries to carry out the collections on their behalf. Individuals or institutions in source countries were paid to collect the specimens but were not offered any share of the benefits that might arise from their future use. In order to retain some sort of record of the number and type of collections that were undertaken within their borders, some states (though by no means all) required collectors to obtain a permit to collect. Bureaucratic vagaries often made the process of obtaining such permits lengthy and arduous, though they cost little, if anything. Although the existence of a permit system implies a commitment to monitor collection activities, this commitment rarely proved to be robust. The principle of "free access" to genetic and biochemical resources was firmly established, and there was consequently little incentive to devote scarce resources to the task of tracking where collected specimens went or how they were subsequently used. Despite the difficulties associated with securing a permit, the permits, and indeed the collections themselves, were rarely examined on departure, resulting in a largely unregulated outflow of genetic resources. As Charles Peters, one of the New York Botanical Garden's most experienced field researchers, noted, "Even though it took like *days* to get these permits you would look *in vain* for anyone at the airport in the source country that wanted to look at your permit. You're like, 'Will somebody *please* look at my permit—it took me three days to get it,' but there was never anyone. You could have left with anything. As far as accountability went, it was like zilch."[2]

This was all to change in the mid-1980s. As reports began to emerge of the potential applications of biotechnology and the range of uses to which genetic and biochemical resources could subsequently be put, countries became more cognizant of their value. They began to demand a right to claim sovereignty over these resources and to secure some form of economic return for their collection and use. Their efforts led to the development of the principles on benefit sharing that were drafted for inclusion in the biodiversity convention. As I noted earlier, the protocols that were introduced under the convention are not legally binding. The bioprospecting industry is a self-regulating one. The terms and conditions under which it operates are set from within the industry and are not subject to review or sanction by any independent regulatory body. Organizations and companies based in signatory countries may feel a moral obligation to introduce appropriate compensatory agreements; those that are based in nonsignatory countries may not. However, while being a signatory to the convention may imply a commitment to the principle of benefit sharing, these commitments are not always translated into practice. Governments do not always place pressure on their nationals to comply; neither do they seek to monitor their activities. Japanese organizations, for example, have undertaken extensive but completely uncompensated bioprospecting operations in Micronesia in recent years, prompting one highly placed international lawyer to remark that as far as he could tell, Japan has no interest whatsoever in the benefit-sharing issue.

Ultimately, responsibility for ensuring that a benefit-sharing agreement is implemented must rest with the government of the source country. If a state requires such an agreement to be implemented before permission to collect is given, it will be—if not, it won't be. Many developing states are only now revising existing legislation to reflect the need for statutes governing access to genetic and biochemical resources and have yet to implement formal policies. Although the United States is not a signatory to the convention, this research suggests that most of the U.S.-based organizations, institutes, universities, and companies that are involved in biological collecting are committed to the principle of benefit sharing. Despite this, only the larger pharmaceutical companies and publicly funded operations (such as the NCI and ICBG programs) have actually introduced formal agreements. Asked whether benefit-sharing agreements were routinely employed within the bioprospecting industry, Steven King, the vice president of Shaman Pharmaceuticals, replied:

> Definitely not. We have gone to some lengths, as you know, to put this three-tiered reciprocity agreement in place to ensure that the process is equitable—but there are certainly a lot of companies out there like [name] and [name] that will supply you with materials for just for a fee—I mean they are fee-for-service providers—you tell them what you're interested in, and if they have it they'll send it to you, or they'll find it for you, no questions asked.[3]

Given that U.S. companies and organizations are not under any legal or even moral obligation to conform to the protocols introduced under the biodiversity convention, it is interesting to consider why they began to develop and introduce benefit-sharing agreements. There were several motivations. Some principled actors within the largest collecting agencies were genuinely concerned to limit the unregulated and uncompensated flow of genetic resources from developing to developed countries; others were worried that a failure to implement such agreements might result in adverse publicity or embargoes on collecting. The first agreement—the NCI's "Letter of Intent" was developed, I believe, in good faith, with the intention of creating a new and more equitable paradigm for governing the exploitation of genetic and biochemical resources. However, had a new compensatory paradigm not been introduced at this time, the NCI's bioprospecting program may well have been disrupted by protests or may have even collapsed. Motivated, then, by a combination of concerns, senior officials at the NCI and the NYBG first began to set about devising a model compensatory agreement in 1987.

It is particularly important to note that this agreement was always designed to provide compensation in phases. The decision to structure it in this way spoke to an awareness that the value of the collected resources would only be revealed once their full potential had been realized and that this would probably occur at a time and in a place far removed from the original point of collection and transaction. It also reflected a tacit acknowledgement that it would be extremely difficult to predict how these resources could be used in the future and thus what their potential value might be. Compensation was phased so that some return would be forthcoming at every stage of the development process. The first phase of (short-term) compensation was to consist of an advance payment (usually a small sum of money) that would be offered to communities in return for a number of collected samples. The second (medium-term) phase would involve providing infrastructure or training for members of supplying communities, while the third (long-term) phase would involve paying collecting institutions or communities in source countries a royalty—a percentage of the profits that could accrue from any products derived from or based on the collected material. This latter element was particularly important, as it allowed new income streams to come online over time, as further applied uses of the collected materials were revealed.

The NCI model was considered innovative, just, and potentially applicable, and it was warmly received by the international collecting community. Within five years of its introduction, agreements closely modeled on this same principle of three-phase compensation began to proliferate, being progressively adopted by collecting institutions and organizations throughout the developed and the developing worlds. The compensatory agreements devised by large pharmaceutical companies such as Merck and Shaman are based on this model, as are those introduced by academic and scientific collecting institutions, such as the University of Illinois and the Royal Botanical Gardens at Kew. The benefit-sharing agreements devised by the consortia involved in the ICBG are also based on the same principle of three-phase compensation. As Gordon Cragg has suggested, the NCI model of compensation has, indeed, turned into something "pretty major." The authors of the original NCI agreement reported in 1997 that it is now employed as a regulatory mechanism in over twenty different countries.[4] Under the terms of the biodiversity convention, each country is encouraged to devise its own regulatory and compensatory agreements so that it may better accommodate political, economic, and cultural variables. Despite this, the NCI agreement and other similar compensatory agreements that are modeled on it, all of which were devised in the United States, have now become so ubiquitous that it is almost impossible to find any collecting institution, in any country, employing any alternative model, as David Downes, senior attorney at the Center for International Environmental Law in Washington, D.C., confirmed:

> Right now I'm helping the secretariat of the biodiversity convention put together a background paper on the use and regulation of genetic resources, and so one thing we're trying to do is to compile a list of interesting different agreements which can illustrate what's going on, and I'm a bit frustrated because so far I haven't been able to find *anything* that doesn't involve the U.S. in terms of the devising of benefit-sharing agreements.[5]

The NCI model for compensating for the use of genetic and biochemical resources provides the foundation for a regulatory regime that has become normative over both space and time, creating, in Featherstone's words, a new "proto-universal culture of regulation." But why and how did this particular regulatory agreement (which is inevitably a product of a contingent set of social and cultural relations) become universalized to the global scale in such a short space of time?

NETWORKS, CAPILLARIES, AND THE GEOGRAPHY OF KNOWLEDGE SYSTEMS

Foucault has argued that in understanding how relations of power operate it is important "to accept that the analysis should not concern itself with regulated and legitimate forms of power in their central locations . . . on the contrary it should be concerned with power at its extremities, in its ultimate destinations, with those points where it becomes capillary, that is, in its more regional and local forms and institutions."[6] It is here, he suggests, that "power surmounts the rules of right which organize and delimit it and extends itself beyond them, invests itself in institutions, becomes embodied in techniques and equips itself with instruments."[7] Such an analysis is certainly central to any understanding of how the power to control trade in bio-informational resources has been exercised in the global economy. For while the protocols introduced under the biodiversity convention provide exactly this sort of centralized "regulated and legitimate form of power" over the terms of use of genetic and biochemical resources, it is in the way that these protocols have subsequently been translated into particular (localized) and culturally informed modes of regulation and then promoted and disseminated, through a linked network of regionally based individuals and institutions, that the key to their current hegemony lies. To understand why U.S.-based compensatory agreements have become so predominant, it is necessary to narrow the focus of investigation—to move attention away from the global scale in order to explore the much more subtle and unobserved negotiations that occurred between prominent actors in the bioprospecting industry, for it was here, at this much more domestic scale, that the parameters of the terms and conditions of this trade in bio-information were first set.

A closer examination of the history of the development of benefit-sharing agreements reveals that they have not, of course, been made by "the United States," but rather by particular institutions within the United States and, perhaps more importantly, by a particular network of actors situated within these institutions.[8] These actors consist principally of a group of corporate executives, lawyers, brokers, scientists, and academics who, despite their widely divergent forms of expertise and differing agendas, have together assumed responsibility for formalizing the relations of exchange that govern this new resource economy in genetic and biochemical materials. Each of the actors within this elite group has highly specialized knowledge of at least one aspect of this trade that he or she has acquired within a specific corporate, disciplinary, or institutional context. These specialized understandings are shared or transmitted between them, and they collectively employ them as the bases upon which decisions about the operation of the trade are defined and negotiated. Although each member of the elite is situated in a particular location—a botanical garden or pharmaceutical corporation for example—the elite itself is only constituted in those instances when the actors are linked together, as Strathern suggests, "electronically, socially or organizationally" to form a wider regulatory network.[9] The "site" that the elite operates in is not a single geographic location but rather the electronic or social space within which the network forms and in which their decisions are transacted and enacted.

Interestingly, although these individuals might characterize themselves as holding an elite position within their organizations, most are only now becoming aware of the fact that they also constitute part of that elite group of decision makers who together control the terms and conditions of exchange for bio-informational resources. In the absence of any more formally constituted regulatory organization, this elite network has acquired de facto but, it would appear, exclusive responsibility for drafting, implementing, and overseeing the operation of policies pertaining to the commercial use and exchange of genetic and biochemical materials and information. This group first began to form in the late 1980s. As I noted earlier, the impetus to develop a compensatory agreement came initially from researchers working in collaborating institutions such as the New York Botanical Garden, who were responding to concerns raised by communities in source countries. It was this group that first approached the NCI and pressed them to develop an agreement. A series of other collaborating institutions that were based in the United States and involved in the NCI collection program (such as the Missouri Botanical Garden), were also consulted in the negotiation of this first agreement.

Most of the negotiations for the development of the "Letter of Intent" took place informally, in a series of meetings between those involved in the organization and day-to-day operation of this trade—researchers, executives, and lawyers working in botanical gardens, pharmaceutical corporations, and academic research institutes. This group then conferred privately with others who were highly placed within relevant nongovernmental organizations and institutions such as conservation, environmental, or legal agencies. It is interesting to note, however, that representatives of affected communities had little *direct* involvement in the negotiation of terms and conditions. Having expressed their concerns to U.S.-based intermediary institutions, their requests were then forwarded on to the NCI by those institutions. Those who were considered by the institutions to be best equipped to act as advocates for indigenous communities were primarily trained in one discipline—ethnobotany. Many were directly involved in the field collection process, as they were often the only individuals based in the United States who had undertaken ethnographic studies on the pharmacopoeia of different indigenous groups. Having acquired knowledge of plant and animal species and of the medicinal uses to which they might be put, along with the trust of local communities, they were ideally positioned to direct the collection process with maximum efficiency. They were also considered

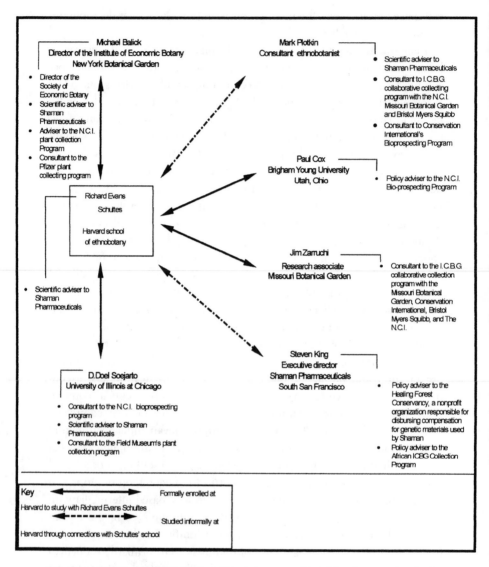

6.1 Schematic diagram of the interconnections between members of the elite network trained at Richard Evans Schultes's Harvard school of ethnobotany.

to be ideally positioned to proffer advice on the development of policies pertaining to the equitable and sustainable use of genetic materials supplied by indigenous communities.

As ethnobotany is a relatively small discipline, it is inevitable that there will be a degree of interconnection between senior practitioners. However, the ethnobotanists who form part of this elite network literally share a common pedigree, being either alumni or the protégés of alumni of the same Harvard University course in ethnobotany that was founded by the renowned ethnobotanist Richard Evans Schultes in the early 1960s (see also Davis, *One River*). By tracing the genealogy of the academic dynasty set out in figure 6.1, it is possible to illustrate both the degree of interconnection between the members of the elite and the pervasiveness of the network that is produced out of these interconnections. As the figure illustrates, Schultes's successors are well placed to play a direct and significant role in the organization and regulation of this trade through a number of different avenues. Several members of this group now have either a direct involvement in corporate enterprises (such as Steven King, the vice president of Shaman Pharmaceuticals) or are actively linked with corporate concerns or conservation groups (such as Michael Balick, who has sat in an advisory capacity on the board of Shaman, and Mark Plotkin, who was vice president of Conservation International). In addition to their principal roles, each member of this network is typically multiply positioned—acting as a director of a scientific or academic collecting institute or company, as a senior policy analyst, and as an adviser to a range of conservation agencies, other collecting institutes, or other pharmaceutical corporations.

What the diagram cannot illustrate is the degree of *informal* collaboration and consultation that takes place between these actors and, consequently, the extent to which decisions about how the trade is or ought to be organized are privately enacted. As well as meeting in official forums, they suggested to me in interview that they also discuss, over the telephone or by e-mail, questions about how effectively contracts work, what constitutes reasonable rates of compensation, and how to counter threatened embargoes on collecting. As Michael Balick has attested, the process of devising appropriate terms for compensatory agreements was largely conducted in this manner:

> Things like the NCI model and the Shaman model have really been developed by a group of us who've sat down and kind of figured it out from the beginning based on our experiences, satisfac-

tions, and dissatisfactions with other areas and other issues and what other companies do and would be best to do. So that was kind of a group thing . . . most of the people—most of the ethnobotanists . . . are in some way or another advising these companies on these issues and have done so from the beginning.[10]

To what extent this degree of collaboration and conferring is attributable to their shared, dynastic, academic connection is difficult to assess, although it is clearly a contributory factor. This is not to imply that the members of the elite have not consulted others in the construction of these agreements. Attempts have been made in recent years to shift the locus of control away from the centers of calculation by involving supplying-country organizations in decision-making processes from which they have traditionally been marginalized. It must be said, however, that this has largely involved consulting with senior administrators working within source country institutions, or representatives of indigenous groups, in order to establish what terms and conditions they would *ideally* like to see included within compensatory agreements. Source country organizations have also drawn upon the pro bono legal expertise offered by public-interest lawyers based in Washington, D.C., to assist them in negotiating contractual and compensatory agreements—as for example InBio did in devising their agreement with Merck.[11]

Despite this, evidence from this research suggests that the terms and conditions of exchange for biological materials continue to be set from within the new centers of calculation on the basis of what is economically viable for corporations (who are the ultimate recipients of the material). As I mentioned earlier, although the United States is not a signatory nation to the biodiversity convention, U.S.-based pharmaceutical companies have been prepared to entertain requests for compensation, if only to protect existing collection programs and avoid adverse publicity. However, they are only prepared to accommodate requests for compensation *up to a point*. As the director of commercial development at Bristol Myers Squibb suggests, this is inevitably determined by the *pharmaceutical industry's* conception of what the material is worth:

> As it turns out, as I'm sure that you are aware, all of these countries use different criteria and have different expectations, and to my thinking it's a big mess. It is very, very difficult to come up to their expectations because they are so out of line with the pharmaceutical industry's ideas of what is and is not possible . . . their expecta-

tions of compensation—the timing of it, total magnitude, types—it was all just very out of line.[12]

Another senior manager at Merck gave a similar response when asked to comment on how corporations determined "appropriate" levels of compensation for collected materials. As he notes, a vacuum was created in the post-free-access era, in which the market value of collected biological materials had yet to be established. In making their assessments, corporations were primarily concerned not to pay levels of compensation that they felt would exceed what was economically viable for the company:

> This is to a large extent a market phenomenon where people who did not attach any value to these raw materials in the past have decided that they have tremendous value and they are seeking to gain tremendous return from these and *organizations such as our own . . . have fairly well defined what value these things have to us* and are simply not willing to pay the excessively inflated prices that are being asked.[13]

The terms and conditions established under the compensatory agreement devised by the NCI were deemed by those actors involved in its development to constitute, *in their collective estimation,* the most favorable that companies and corporations could agree to. As this is a market economy in genetic and biochemical materials and as companies and corporations are the principal end-point consumers of these resources, it is should be unsurprising to discover that they ultimately determine the materials' commercial worth.

The NCI agreement was promoted and disseminated initially through the NYBG's network of collaborating institutions in source countries and through the NYBG's institutional and corporate linkages with other organizations such as Merck, the University of Illinois, the Missouri Botanical Garden, the Royal Botanical Gardens at Kew, and their collaborating institutions worldwide. Institutions in source countries were of course not obliged to accept the terms and conditions of this agreement; they have always been at liberty to devise alternatives. However, as Lisa Famolare of Conservation International suggests, most have had neither the expertise, money, nor opportunity to undertake such an exercise:

> The fact of the matter is that, really and truly, people don't have the legal resources to re-create these things [compensatory agreements], and once somebody says "here's a model" as they did in the

bioprospecting book[14]—I think most people are using something very similar to that or to NCI's model. Missouri I think is, New York Botanical Garden is, I don't think they're that different.[15]

It is through these processes that the NCI's three-phase model of compensation has become universalized over space and time and, in the process, institutionalized as an appropriate and effective means of compensating for the use of genetic and biochemical resources. Given that this model agreement is now utilized all over the world and has become the one principal mechanism through which the collection, utilization, and compensation for use of genetic and biochemical materials is now regulated, it becomes enormously important that the agreement deliver on its promise to provide a just and equitable means of governing this nascent, but increasingly lucrative trade. In order to assess whether it has (or indeed can) fulfill its promise, it is necessary to examine how such agreements operate.

COMPENSATORY AGREEMENTS:
INVESTIGATING TERMS AND CONDITIONS

Although the three-phase NCI model compensatory agreement has been widely adopted within the bioprospecting industry, other types of benefit-sharing agreements that are based on just one or two stages of compensation do exist. Given that there is clearly some public-relations mileage to be made out of having a compensatory agreement in place, it is not surprising to find that many companies profess to have implemented some sort of benefit-sharing agreement. However, as we are emerging out of what was an era of entirely unregulated and uncompensated access to genetic resources, it is also unsurprising to find that many companies and organizations have very modest conceptions of what constitutes "benefit sharing." For some, this simply consists of paying out, in advance, a fee per sample of collected biological material. The amount paid per sample varies, but according to various industry sources usually ranges between $25 and $200 dollars per dry-weight kilogram of collected material, with an average falling between $50 and $100 dollars. Microbial cultures cost between $20 and $140 dollars per sample. Payments for prepared biochemical extracts range between $100 and $200 dollars per twenty-five-gram sample.[16] While these payments could conceivably be argued to constitute "compensation," in fact they barely cover the costs source-country organizations incur in collecting the material, which, as Laird notes, include "the cost of shipping the specimens from the field to the collecting institution, and then on to the company, confirming initial field identifications, curating voucher specimens, processing data, conducting literature searches and maintaining staff and administrative infrastructure."[17] Despite this, many smaller companies make no further commitment to benefits sharing.

The most problematic aspect of these basic "fee-per-sample" compensatory agreements is that they are based on the same principle as traditional resource-management regimes designed to compensate for the use of fixed amounts of stock resources. These regimes presume a cycle of consumption and resupply. They are based on the principle that the recipients of the resource will be provided with an initial supply of material, for which they will pay a fixed amount, but that they will in time consume that material and return for a further supply, at which point the supplier will secure further income or the opportunity to renegotiate existing contracts on more favorable terms. However, as the evidence produced in chapter 4 suggests, the probability of companies returning to secure ongoing supplies of genetic or biochemical materials is rapidly declining. The creation of new technologies that enable companies to extract, store, and then replicate these materials over both space and time, and the creation of new networks of trade and exchange that provide alternative sources of supply for such materials, are obviating the need for recollection. Senior players within the bioprospecting industry are acutely conscious that these developments fundamentally alter the social and spatial dynamics of collection, a fact that, as Michael Balick of the NYBG has acknowledged, ought to be reflected in the structuring of compensatory agreements:

> Sure, it's very different now because all you need is a bit of the plant. You only have to go there one time, and then it's over as far as control of that resource and access goes for the people who own it, so the deal has to be brokered with them in mind from day one. So absolutely it's not like you have to go back to the village to keep filling up your truck with coal; you have the plant and that's it.[18]

More sophisticated agreements, such as that devised by the NCI, offer compensation in stages in recognition of the fact that these biological resources are now likely to be used successively, in much the same way that other informational resources are, rather than being simply consumed as a material resource as they often have been in the past. They usually provide an advance payment, infrastructural support and training, and a royalty on products produced from collected material. The up-front payment made under these agreements should not be compared to fee-per-sample payments made under other agreements, as it forms part of a total compensatory package. The amount paid in advance may consequently vary in accordance with the weighting given to the other elements of compensation. A large advance payment may result in a reduced royalty rate; conversely, some suppliers may forgo an advance payment altogether in the hope of securing a higher royalty rate.

Interestingly, the weighting that is given to advance payments within many three-phase compensatory agreements appears to be changing. When the Merck–InBio agreement was signed in 1993, the up-front payment was $1,135,000—an unprecedented sum that has certainly not been rivaled since. The size of this payment provided the focal point for much of the favorable publicity surrounding the launch of the agreement, despite the fact that its purpose was to defray the costs InBio would incur in undertaking Merck's two-year collection and sampling program. The payment did, however, form part of a larger package of compensation which included a provision for training and infrastructural support and a royalty percentage of between 2 and 5 percent of net profits on products created from the collected materials.

Although three-phase agreements are still widely employed, industry support for the principle of paying compensation *in advance* for the collection of material appears to be waning. While, for example, all five of the new ICBG agreements include some form of staged compensation, only two include an up-front payment. This change is significant, for, as the co-coordinator of the program, Dr. Josh Rosenthal, of the NIH's Fogarty International Center, has suggested, the purpose of these payments is both functional and symbolic. They were initially included to provide both an immediate form of reciprocity to supplying communities as well as "evidence of the sincerity and commitments of the partnerships."[19] Advance payments, when made, are normally paid into a trust fund to which local communities may apply for small grants for urgent needs. However, as Rosenthal suggests, corporate partners are increasingly unwilling to make up-front payments. As he ruefully admitted when interviewed on the progress of the ICBG program,

> there aren't any up front monies yet; they're still tying to get them out of the pharmaceutical partner. I always find it difficult. . . . I mean you need to set up trust funds, and it is hard to know how that is going to happen in advance if there's no [money]. . . . It's a lot

better if there is actually a chunk of money to make it a real thing. So I'm suggesting to them [the leaders of each consortia] that they should go back to the pharmaceutical company again and try and get some up-front money from them.[20]

The disinclination to make advance payments is driven, in large part, by the fact that companies are reluctant to pay compensation for biological materials that may prove to have no immediate or ultimate use. Although executives within pharmaceutical companies acknowledge that genetic and biological materials form the basis of many valuable pharmaceuticals, many hold the view that "raw" samples constitute an entirely undeveloped resource with little inherent value. As Dianne De Furia at BMS argues,

> raw samples of plant material are so far from a marketable product that you cannot value them at anything more than ten dollars a sample or something like that. . . . If you've demonstrated to me that this is a significant medicament for some sort of infection, ok, well, we still don't know how to produce it, we don't know what else can be done with it, certainly it's not developed, and it's not marketed. Well alright, *maybe* I'd be willing to say, on the basis of the preliminary data you have shown me, that I will give you a 5 percent royalty on the end product plus several hundreds or thousands or millions as we go, *if* it proves itself worthwhile. But when you ask yourself, "What am I willing to promise somebody to go out in the woods and pick something?—What are we willing to promise for something that might come out of that?" Well who is putting in all the effort and where does all the investment come from? It's not from that individual [collector], it's not from that intermediate broker either. So those numbers [payments] would be, justifiably, very, very small.[21]

Consequently, within the ICBG agreements, comparatively more emphasis has been placed on medium- and longer-term forms of compensation, including royalty payments. The concept of paying a royalty can be perceived as an enlightened measure introduced (as it has been in other fields) to both acknowledge and compensate for the successive uses made of particular materials or information over both space and time. While this remains the principal rationale for introducing such a measure, it is also important to note that a commitment to pay a royalty can also effectively absolve companies of the obligation to pay compensation for anything other than those few samples that they attest have been of use to them. The question of how it is possible to establish with any certainty which have been of use will be discussed shortly. Before turning to this question however, it is important to discuss briefly the role of what are referred to as "medium-term, nonmonetary" forms of compensation.

INFRASTRUCTURAL SUPPORT AND TECHNICAL TRAINING

Many of the companies and institutions that collect samples of genetic and biochemical materials have agreed, as part of their commitment to phased compensation, to provide suppliers with medium-term benefits. Most determine what form this compensation will take in consultation with organizations and institutions in the source country or through government agencies. The NCI agreements, the Merck–InBio agreement and the ICBG agreements have been negotiated in this fashion. Other companies have decided to address this issue through different means. Shaman Pharmaceuticals, for example, elected to create their own nonprofit organization, called the "Healing Forest Conservancy." This organization was established, in the words of its founder, "because no nongovernmental organization existed to provide a formal and consistent process to compensate countries and communities for . . . the commercial development of biotic products."[22] The role of determining and disbursing compensation has now been assumed by the "independent board of directors and advisers" who oversee the work of the conservancy.[23]

It is important to briefly consider the role of the state in these compensatory processes. Interestingly, although under the biodiversity convention genetic and biochemical materials became formally constituted as part of the natural patrimony of each nation-state, government departments do not always play a particularly active role in determining appropriate forms of compensation or avenues of disbursement.[24] Responsibility for these processes is often devolved to or assumed by those government-funded institutions in source countries (such as herbaria, museums of natural history, or university departments) that are directly involved in the process of collecting material and with whom collecting contracts are let. Consequently, while medium-term benefits could theoretically take any number of forms, from legal assistance for land rights claims to literacy classes or new public facilities, they usually take the form of training and infrastructural support for scientific institutions concerned with the identification, use, and management of biodiversity.

Table 6.1   InBio's use of Merck's $1,135,000 payment

Contribution to the Ministry of Natural Resource's National Park Fund	$100,000
Training for Costa Rican scientists	$120,000
Payment to the University of Costa Rica for extracting samples	$80,000
Equipment for chemical extraction at the University of Costa Rica	$135,000
Salaries for chemists, laboratory assistants, and so on	$100,000
Contributions to parataxonomists' work on the national biodiversity inventory	$60,000
Automobiles, fuel, oil, laboratory supplies, field supplies, per diem expenses, and so on for collectors	$120,000
Equipment for the biodiversity inventory: hardware, cabinets, solvents, computers, and so on	$285,000
Administration and overheads	$135,000
Total	$1,135,000

Source: A. Sittenfield, quoted in S. Laird, "Contracts for Biodiversity Prospecting," 110.

There is an argument that these forms of infrastructural support and training provide a much needed mechanism to improve research and development capacities in source countries, to identify, assess, and monitor regions of high biodiversity, and to enhance databases of information and inventories of existing species. The question remains however: *Who benefits from these improved capacities?* Under the Merck–InBio agreement, for example, almost all of the advance payment that InBio received—in excess of one million dollars—was expended providing InBio with the training and equipment necessary to access, identify, classify, and collect biological materials and to create biochemical extracts from them on Merck's behalf (see table 6.1). This has been the case in many other agreements.

A close examination of the details of this phase of the compensatory package reveals that "training" usually comprises the acquisition of skills in "plant collection and drying in the field, extraction, testing, compound isolation, identification and modification, databases development and management, . . . use of Geographic Information Systems, . . . understanding of intellectual property rights, ethnobotany, . . . curatorial methods, . . . and commercial production of medicinal plants,"[25] while "infrastructure" includes milling machines, freeze dryers, chemicals, microscopes, plant presses, cell-line equipment, glassware, computer and GIS software, bar-code readers, fume, hoods, vehicles, and the like.[26] The recipients are either local workers, who are trained to form a taskforce of "parataxonomists" capable of identifying and classifying specimens in the field, or scientists and technicians who are sent to U.S. laboratories, research institutes, or universities for further specialist training. As Rosenthal notes, over 1,400 individuals have received some formal training in these specific skills through the ICBG programs. However, the vast majority of these have participated in short one-to-seven-day or three-to-six-month courses; only eighty people having been enrolled in a higher-education degree or course. Training and infrastructural support of this particular kind has formed the principal component of medium-term benefit sharing in all of the NCI's compensatory agreements, in those that have been offered by companies such as Bristol Myers Squibb, and in all of the ICBG agreements.[27]

The emphasis that is placed on these particular forms of training and support, to the exclusion of all others, can be construed in two ways. One frequently promulgated argument is that compensation *must* take this form if developing countries are to acquire the capacity to turn basic raw materials into exportable "value-added" products and in so doing secure the means to break the existing cycle of resource exploitation. This is undoubtedly true; however, the task of achieving independence and competitiveness in production is an inevitably complex one. For it is by no means clear that the type of infrastructure and training that is offered under these agreements serves to do more (in the short term at least) than turn source-country institutions into ever more efficient mechanisms of extraction. Although they can be seen as providing a first step to self-sufficiency in drug production, the compensations that are currently offered—training programs in specimen identification and collection, the provision of equipment and vehicles for undertaking collections, the renovation and improvement of laboratories and storage facilities, and even the supply of geographic information systems—must be viewed first and foremost as providing a necessary means of improving the speed and efficiency of the extraction process. They do so simply by enabling materials to be identified, collected, and transported in a more cost-effective manner.

Supplying countries achieve a marginally higher return for an extract or cell line than they do from the sale of whole organisms or bulk samples; however, the real pecuniary advantage accrues primarily to the collection agency. As I have illustrated, it is much more efficient for collecting organizations and institutions to transport and store collected materials in a proxy form, as a cell line or biochemical extract, than as whole organisms or a bulk sample. It surely cannot be coincidental that most of the training that source country scientists and technicians receive is designed to improve their ability to create just such proxies. As Laird noted when commenting on Merck's decision to build extraction and screening facilities in Costa Rica, "it is easier and cheaper to transport and import extracts than raw samples, and building research capability at InBio and the University of Costa Rica will still cost Merck less than employing similar resources in the United States. Merck supplied chemical extraction equipment to the University of Costa Rica but has exclusive commercial use of these facilities."[28] Even the equipment and training that source-country institutions receive in taxonomy, GIS, and database development for biodiversity management and conservation work provide advantages for bioprospectors. While all undoubtedly assist in efforts to improve approaches to ecological management, they are also used, as Rosenthal et al. have noted, to "help to identify areas and periods of high biodiversity for collection."[29]

Few attempts have yet been made to fully quantify the contribution that these medium-term forms of compensation—infrastructural support and training—actually make to capacity building in developing countries, and a more detailed audit may, in time, reveal it to be more substantial than it now appears. Some recent preliminary research undertaken by Carolyn Crook, a geographer from the University of Toronto, suggests, however, that while the Merck–Inbio agreement (which has since been renewed several times) has generated further revenues, they do not rival those that would otherwise have been created from activities such as forestry or ecotourism and have not been shared with local or indigenous communities "to any great extent."[30] As she notes: "compared to alternative values of the land, for example in timber values alone, the earnings appear unlikely to [greatly] increase the value of natural ecosystems."[31] She is, however, of the opinion that the investments that have been made in Inbio have increased research capacity to some degree and notes that this has resulted in more detailed investigations of diseases such as malaria and of agricultural problems, both of which remain of little interest to the multinationals.

Medium-term compensation may occasionally take the form of humanitarian aid—the provision of health care or medications or the construction of schools or other forms of public infrastructure, but this forms a comparatively small part of most compensatory packages. The Surinamese ICBG has, for example, provided money from up-front payments dispersed to its Forest People's Trust Fund for the provision of improved local transport networks and small-scale community sewing and agricultural projects, while other ICBG programs have provided some funds for health care and primary schooling.[32] It would be overly cynical to suggest that companies or organizations provide these compensations because they are likely to benefit them in some way, for in most instances they do not.

There are, however, some instances in which they do. In 1992, for example, Shaman Pharmaceuticals began a research expedition to the Ecuadorian Amazon. As part of their stated commitment to benefit sharing, the company entered into negotiations with tribal village leaders to establish what form they wished the compensation to take. They responded with a request for a reticulated water system for the village. Initially, Shaman agreed to this request but later reneged on the agreement. Steven King explained why this decision was taken and what alternative was offered:

> The cost involved in supplying water to all the houses in the village was just too prohibitive. So we discussed the matter with them again and it was decided that the money would be put to better use by building an extension to their airstrip. You see, at the time, it was only long enough to get a two-seater plane out, and by lengthening it we could get a bigger plane in there which would mean that they could have someone, like a relative, accompany them if they have to be flown out to hospital.[33]

Perhaps unsurprisingly, King confirmed that the extension to the airstrip also provided efficiency gains for the company by improving access to the collection site and increasing the size of the planes that could be used to transport collected materials out of the area. King insisted that this was not the principal rationale for choosing this option, and that may be so. It was, however, a fortuitous consequence.

Although it is difficult to obtain clear information about how much money companies expend on medium-term forms of compensation in total, it is possible to arrive at an indication of their annual expenditure through a search of the literature. Shaman, which is considered within the industry to have one of the best records for providing monies for training, cultural and social development, and infrastructure, spent $1,500 on the purchase of materials and labor for the extension to the airstrip, $11,000 on the construction of a school in Thailand, and some $11,000 training women in species identification and collection in Lucknow, India, in one year. They also conducted some medical clinics and provided some antimalarial medications in areas in which they collect. Whilst these contributions are undoubtedly important, they represent a tiny portion of Shaman's overall research and development budget, which rose from $5.4 million in 1992, to $13.6 million in 1993, to $18.4 million in 1994.[34] Similarly, while much was made publicly of the million-plus dollars that Merck provided to InBio for training and infrastructure, this money represented only a tiny percentage of Merck's total of more than $1 billion research and development budget for the fiscal year of 1992 to 1993.[35]

It has been argued that training programs and scientific infrastructure are important as they enable source countries to make more accurate inventories of their natural resources and create their own collections of genetic and biochemical materials. In fact, under the protocols established by the biodiversity convention, bioprospectors are required to deposit a duplicate collection of any material that they identify and remove in a designated institution in the supplying country. However, while provision of taxonomic identification programs, training programs, and basic scientific infrastructure helps countries to create their own collections of material, it does little to help them convert these collections to novel and patentable pharmaceuticals. Achieving this task demands not only concentrations of materials and basic training but also access to sophisticated technologies and advanced scientific expertise. While some laboratories in the South are acquiring these advanced capabilities, they remain few in number. At present, most of these very advanced technologies remain concentrated in what I have described here as the "new centers of calculation." As one senior corporate chemist put it, "what we've got here [at Merck] is a situation where we've got the best taxonomists, the best technologies and the best collections *together*. . . . You need all of them to make the link. Having a library full of [plant genetic] information is great, but somebody has to know how to read the books and know what to do with the information, and that's what we have here."[36]

Without access to this very advanced technology and expertise, the collections of plant material will remain, as Borris suggests, unreadable books, comparatively valueless material resources. As Franchesca Grifo has suggested, it is clear that there must be a commitment on the part of collecting agencies to transfer more advanced drug development technologies to source countries, in part, as she put it, "because that is the only thing that's going to get us away from the neocolonial aspects of it—otherwise it's no different."[37] However, when asked to estimate how many collecting agencies were actively committed to this objective, she regretfully replied "almost none. . . . I think the ICBG's are pretty rare in that. Shaman, for example, are going back and doing things, but that's not their goal, for people to be able to do this independently of the U.S. Their goal, in my opinion anyway, is more humanitarian, you know, working towards improving health conditions, cultural and other kinds of projects."[38]

As Grifo notes, it would be unusual indeed to find many corporations or companies that would feel an obligation to fully share their advantages (superior technologies and expertise) with their competitors. This is understandable; however, without such commitments, developing countries will remain confined to their current role of supplying semirefined raw materials. Recipient companies will, meanwhile, be able to use the technologies there concentrated, to isolate, manipulate, combine, and even replicate collected genetic and biochemical resources to produce new combinations of material and information that can be employed to advantage, over and over again. It is only through these ongoing processes of exploration and exploitation of the material that the full value of the resources will be realized.

FUTURE BENEFITS: ROYALTY PAYMENTS

While some supplying countries have secured some up-front benefits in return for the collection and utilization of their genetic resources, these have not usually been substantial. The justification for keeping up-front benefits at a minimum is that further benefits will be forthcoming when and if these samples prove to have some utility. This is based on a presumption, first, that companies will return to the source country for bulk supplies of the material once they move into commercial production and, second, that the provision of a royalty payment will see a percentage of the future profits that accrue from products developed from collected material returned to the source country. There are, however, several difficulties with these presumptions. As the evidence produced here suggests, the probability of companies returning to source countries for ongoing supplies of material is not great. Although those who are involved directly in the collection process continue to suggest that collectors will return to source countries to grow or gather more organisms, those with experience and responsibility for negotiating contractual agreements governing the disbursement of benefits have a more pragmatic view. As Kate Duffy Mazan, a senior attorney with the NCI's Office of Technology Transfer noted in interview, "we do have some cultivation projects going on in a couple of countries—that is one of the things that we urge them to think about, but in truth, when you are dealing with natural products they usually end up synthesizing it, going to a synthetic formulation."[39]

A number of principled actors within the industry have recognized this and argued that in order for the process of exploitation to be equitable, a royalty must be paid to suppliers of material whenever that material is used as the basis of a commercial product, even in instances when the product is ultimately produced by replicating, modifying, or synthesizing the original material. The decision to include a royalty payment among compensatory measures is a development of the greatest significance and one that represents a sea change in the way in which biological materials are both understood and valued as resources. For although systems of intellectual property rights law, such as Plant Breeder's Rights, have provided protection for the successive uses that are made of whole organisms, this is the first occasion in the history of natural-resource governance that a provision has been made to compensate for the successive uses that are made of the genetic and biochemical components of those organisms. The fact that the mechanism that has been adopted to meet this requirement (the royalty) is also used to compensate for the successive uses that are made of other types of information-based products, such as literary works, musical compositions, and computer software, is particularly significant. It suggests that, as with these other information-based products, what is of greatest value and requisite of most protection is no longer *the form* of the work but the *transmissible content* of that work.

Although based on similar principles to copyright regulations, the rate of royalty that is paid to the suppliers of genetic materials rarely rivals that paid to other suppliers of information-based products. The creators of software packages, authors, and musicians would expect to receive something in the order of 10–15 percent of net profit on the successive uses that are made of their works. The industry standard rate of royalty offered to the suppliers of genetic materials is 1–2 percent for a sample of biological material and 3–5 percent for a prepared extract. Rumors of "royalty rates" of 50–60 percent arise out of confusion about royalties and the disbursement of royalties. When reference is made to one partner in a consortium receiving 60 percent royalties and another 40 percent royalties, this is not 60 percent and 40 percent of total net profits—it is 60 percent and 40 percent of the total royalty rate, which is usually between 1 and 5 percent of net profits.

The key to understanding the discrepancy between the royalty rates paid in these respective industries lies in questions of "authorship," "contribution," and "invention." The rate of royalty paid to the producers of musical works, computer software programs, literary works, and so forth are higher in recognition of the investment of intellectual energy and creativity required to author such works. Biological materials are not considered to be an "authored work," and the suppliers of these materials do not, therefore, receive these higher rates of royalty. However, as I have illustrated here, biological materials may be engineered in a variety of ways such that they are transformed by the application of human ingenuity. In such cases, the new "work" is considered to be an invention to which patent rights may be sought. Although it is often assumed that this engineering results from the application of sophisticated technologies, it is evident that biological materials have also been transformed, in important ways, through the application of knowledge-based, so-called low-technologies such as systematic interbreeding or applied intergenerational use. These practices have also produced new "inventions," such as new strains of crops, animal breeds, or medicaments.

As many other writers have noted, the application of generations of indigenous knowledge about the therapeutic uses of particular natural materials (methods of application, suitability, and dosage rates of plants, fungi, venoms, and so on) adds considerable value to what may otherwise appear to be an unprocessed natural material.[40] These groups or individuals could thus be said to have "authored" or "invented" a new use for an existing material, even though their contribution may not be accredited as such. A long and intense debate has taken place within the bioprospecting industry about how best to recognize and compensate for the intellectual contributions made by indigenous groups in identifying and refining applied medicinal uses of particular natural materials, particularly in circumstances where that information has been used to direct collection activities more effectively. It has been agreed (in principle at least) that royalty rates should ideally rise to between 10 and 15 percent (the equivalent to rates paid to authors in other related industries) in instances where a supplying individual or community provides material with confirmed efficacy against a specified disease state. However, as Gordon Cragg of the NCI has noted, although a higher royalty should be paid, it is not always forthcoming. As he explained:

> The highest royalties would come when someone [a supplying individual such as an indigenous healer] who knows, says this is good for this, and drug company A says "by golly it *is* good for that"— unadulterated and unchanged—that *should* be a pretty high royalty. What are the numbers that ought to be attached to that? . . . Who knows? They are probably lower than they should be. If you think about things like films and written material—you know author royalties and the like . . . it's the same sort of thing.[41]

Although royalty rates are not usually remotely comparable to those paid in other information-based industries, they could, nonetheless, theoretically provide a substantial return for suppliers. As Bob Borris at Merck has noted, "when you turn around to the other side of the equation and look at the profit potential for a major product, for example Methaclor, which is a fungal product . . . this is a product that sells better than a billion dollars worth of pills a year, so you're talking about a substantial amount of money, if you are talking about a 1 percent royalty on that you are still talking about a lot of money."[42] It is crucially important to note, however, that as I write this, more than a decade after the first major collecting programs of the modern era were instituted, to my knowledge all royalty payments still remain projected; nothing has yet been disbursed to supplying countries in recompense for the use of collected materials through this mechanism. The explanations that are offered by those working in the bioprospecting industry for why this is so are, in my view, curiously contradictory. The most common justification (and one recited routinely in every recent analysis of bioprospecting activity) is that the process of drug development is a very drawn out, and that it may, thus, take ten to fifteen years before a product reaches the market and profits are generated and disbursed.[43]

The promise of royalties was initially widely used as an incentive to encourage countries and communities to allow collections to occur within their borders. However, more recently (now that many collections have taken place), collectors have begun to switch their emphasis, warning their operatives that they must "be very clear with host country collaborators and regulators about the low probability of major pharmaceutical drugs emerging from any one project, [and therefore] to do everything they can, to provide *tangible near-term benefits*."[44] This echoes a much-repeated wider argument that few, if any, of the materials that are being collected will ultimately be used in the development of marketable pharmaceuticals. Yet how are we to reconcile this argument with the also oft-quoted (and proven) statistic that 25 percent of all prescription drugs are derived from natural materials? Are we to understand that natural materials are suddenly and inexplicably ceasing to be a crucial component of pharmaceuticals? This seems unlikely. Or are we to imagine that although the vast majority of the natural material that is collected for use in the pharmaceutical industry is collected in developing countries, it is somehow not *this* material that will form the basis of valuable drugs and therapies? This also seems unlikely.

A more rational explanation for the apparent lack of products and, hence, royalties may lie elsewhere — in the changing ways these natural materials are now utilized in the pharmaceutical industry. Collected biological materials may well have formed the basis of new pharmaceuticals and therapies over the past decade, but how would suppliers know if they had?

While royalties theoretically allow suppliers to receive a return each time collected materials are used, this return will only be realized if three inter-related conditions are met. First, the collected sample must form the basis of a marketable drug. Second, companies or organizations must *acknowledge* that a collected sample of material from a given location has formed the basis of a marketable drug and that the material has not been acquired from some other nonrecompensable source—such as ex situ collection or domestic cultivation. Third, successive users of the material must be committed to, and capable of, distinguishing where the genetic and biochemical resources that they are utilizing came from. In other words, each must be prepared to establish the "provenance" of the material, for it is clearly impossible to pay a royalty unless it is possible to determine to whom that royalty should be paid. Collected materials are now subjected to numerous processes of modification, synthesis, and replication; they are circulated to and utilized by a variety of different recipients and, in addition, may be cryogenically stored and subject to similar processes of transfer and manipulation in the future. Even if the desire to trace the provenance of the material is there, it may simply prove to be technically impossible to keep track of these materials as they percolate through these innumerable processes of recombination and recirculation.

As practitioners of information technology law have discovered,[45] the task of tracing the successive uses that are made of particular information-based products over both space and time is being greatly exacerbated by the fact that new technologies are having the effect, to use their words, of "dematerializing" products such as books which previously only enjoyed a "hard-copy" existence.[46] This process of dematerialization, they argue, greatly increases the ease with which the information contained within the book can be transmitted or conveyed, circulated among users, copied (often in an unauthorized fashion), or sampled. Although it has yet to be fully recognized, a similar process of decorporealization, is, in this case, also undermining attempts to keep track of the successive uses that are being made of genetic and biochemical materials and information. Many commentators within the bioprospecting industry continue to assert with confidence that "proprietary arrangements can ensure that compensation from commercialization of natural compounds or synthesized chemicals revert fairly to the peoples and nations of original production."[47] However, it seems apparent that the dynamic and mutable nature of this complex resource may well defy attempts to effectively control or regulate its flow. Some examples may serve to illustrate the point.

TAMING THE SLIPPERY BEAST

The difficulties of tracing the uses that are made of collected samples of genetic or biochemical materials are both spatial and temporal. The spatial difficulties arise at a number of scales—global, national, regional, even microscopic. Although nation-states or communities are entitled to receive royalties or other benefits on products based on materials collected within their borders, they will be unable to further these claims unless they are able to prove that the material was actually sourced from their country. This can prove a difficult task for several reasons. Plants and animals rarely respect national boundaries, and many are endemic to a comparatively wide geographic region. The existence of a collecting contract ought to ensure that benefits revert fairly to suppliers—as long as the contract is honored. This decision rests entirely with the collector. As the evidence presented here suggests, companies now collect in many countries, and it may well be impossible for any source country to prove retrospectively that the material that formed the basis of a successful drug was, in fact, sourced from within its borders. Collectors may secure an initial supply of material in one country but when asked to provide benefits may argue that they, in fact, collected them in another. Attempts by countries to argue for benefits on the basis of endemism fail more frequently than they succeed, as Steven King revealed in an interview:

> I just went through the Philippines and they wanted us to agree that if we developed something from plants that were endemic to the Philippines that they would have the freedom to market them in their country—that is to say, to license it in their country and not to have to pay for it—on the basis that if it only occurs in their country, then it's theirs. The reality is that you may have an endemic species in the Philippines and a related species in Borneo or some place nearby and different species which are even more widespread, or common, all of which have the same compounds as the endemic species. . . . The distinction sounds nice, but when you really come down to it, you can't guarantee that what you got there is unique and novel to that country. I know that is something that more and more people are trying to define, but it's very difficult because nature doesn't really comply with that distinction, and so we really couldn't agree to that.[48]

Many countries also have to contend with the unauthorized collection and export of natural materials, a historical problem, but one that has been greatly exacerbated by the fact the most valuable components of organisms (the genetic and biochemical resources embodied within them) can now be conveyed independently of them, in a partially or even wholly decorporealized form. Where border controllers once had to search for whole organisms or collections of organisms, they must now detect the illegal export of tiny samples of material—a development that has made the process of regulating the movement of such resources infinitely more difficult, as King confirmed:

> The fact that they now have to take so little means that issues of control and regulation are incredibly difficult. I mean, people will risk their lives for a pocketful of heroin! We were in Indonesia when [name] heard four Germans in Irian Jaya discussing how they were going to smuggle out medicinal plant samples taking them via Japan and via the U.S., so I mean it is going to be very difficult. Even tracking the materials in a straight vertical fashion is going to be very difficult. I mean, if there's money in it and people are desperate enough, they'll do anything. As someone pointed out to me the other day, people are still smuggling drugs in and out of the Philippines like crazy, and that carries the death penalty—I mean, do you think people have got any chance of controlling the movement of milligrams of genetic material?[49]

It is precisely because it is so difficult to trace how much genetic material is being smuggled out of countries each year that it becomes virtually impossible to ascertain the true extent of this illicit trade. It seems reasonable to assume, however, that it may well increase, as particular species (including some human racial groups) become increasingly rare and as bureaucratic requirements and constraints act to complicate and restrict collecting activities. It is not only genetic materials that are being exported without recompense. As ten Kate and Laird note, indigenous information about the use of collected materials has also become "de-coupled" from the communities in which that knowledge was generated. The information is increasingly stored on databases and found in literature that is published online. When rendered in this form, the information can be accessed through the Internet—obviating the need for interested parties to travel to such communities or, indeed, to pay for access to the information.[50]

The advent of microsourcing creates similar dilemmas, although here the spatial issues are compounded by temporal ones. Replicable quantities of genetic and biochemical material are now being secured through the "remining" of existing scientific and academic collections and then recirculated to industrial users in various locations. As these ex situ collections stand outside the remit of the biodiversity convention, those who access the materials for commercial purposes can use them while avoiding any obligation to share benefits with the original suppliers of the material. As David Downes of the Center for International Environmental Law has confirmed in an interview, materials that were collected in previous decades for purely scientific or academic purposes are now being traded internationally in what is, effectively, a black market for genetic and biochemical resources. As he explained, the probability that this market can be effectively regulated or even monitored is exceptionally low:

> I think the biodiversity convention can be a framework for working out what the basic standards are, but as for enforcing them—well look at the software industry! As you rightly pointed out, I think that is going to be really hard because it's really hard to track this stuff and, as the technology evolves, it will probably only get harder. It's here already actually. Screening technologies exist now that will allow people to take samples in museums and herbariums—things that have been sitting around for a long time—and take tiny little biopsies from the samples and get commercially useful information from them about their genetic structure, and when you get to that stage . . . how much leverage does the source country have? There are huge collections already extant in London, Washington, and many other places. I read recently that this large Japanese marine biotech institute, [name], has been talking with Russia about—there is a huge collection of marine organisms in Vladivostok—and they are trying to negotiate access to that collection for the biochemicals. I mean, why go to the source when you can use this collection? And how do you enforce the collecting principles in that situation?[51]

As improving technology turns existing collections of material from taxonomic curiosities into viable sources of industrial raw material, attention has turned to the question of whether it might be appropriate to extend compensatory mechanisms to cover the use of archived materials in ex situ collections. Even if this were possible or desirable, many would still object to the materials' being used for commercial purposes. They would argue that it would be impossible to do this ethically, as users could not obtain the informed consent of the suppliers of those materials for the proposed change of use (as most are long dead) or determine to whom any royalty should now be paid.

The question of how compensation for the prospective use of collections that are being created now and cryogenically stored for future use might be determined or disbursed also awaits effective resolution. Although agreements that are signed now may include a commitment to pay compensation for the successive uses that are made of collected material in the future, how will it be possible to enforce this commitment over time? Henry Shands, director of genetic resources at the USDA, confirms the problem:

> That's one of the real problems that people haven't addressed because things can sit in a gene bank for a hundred years and of course that exceeds the lifetime of the corporate history—there is no history left then unless it's recorded on a database, so we're really starting from scratch with that kind of thing. By that time the whole IPR structure will have changed, and you'll probably have a whole new set of issues—that might not be such a bad thing![52]

Although the difficulties of tracing the whereabouts of collected materials as they circulate between users at a regional, national, or global scale are well documented here, they pale into insignificance when compared to the task of attempting to track the microscale journeys that these materials undertake as they are moved about and combined within the laboratory. Some collectors have agreed to pay a royalty on the successive uses that are made of particular collected materials, but how will it be possible for them to honor this commitment when it may prove technically impossible to trace the movement of particular gene fragments and molecular compounds as they progress through the myriad cycles of manipulation, combination, and recombination that form the basis of so much of contemporary drug design and manufacture? This task has, of course, become infinitely more complex now that it has become possible, indeed more cost effective, to synthesize or clone the genetic or biochemical materials that are used in the production of pharmaceuticals or gene therapies. Neil Belson, the president of Pharmacognetics explained:

> You know they are extremely unlikely to use a natural compound in a pharmaceutical. What they want is leads—they want to know that a structure is really effective . . . and then they are going to synthesize something based on that natural structure that is easier to produce and maximizes that activity and minimizes the toxicity. So the chances of that natural compound going into a commercial product in its natural form are very, very small.[53]

Given that the greatest value to be obtained from the use of biological materials is increasingly going to be realized by utilizing and reutilizing this genetic or biochemical "information," it is essential, if this new phase of resource extraction is not to be unfavorably paralleled with earlier colonialist exercises, that existing mechanisms be capable of tracing the flow of this bio-information as it is circulated around the many new networks of exchange that characterize this burgeoning trade. Although they rarely discuss them, it is clear that even the most senior practitioners in the bio-prospecting industry harbor some serious doubts about the effectiveness of these mechanisms.

REGULATING THE UNLICENSED COPYING OF BIO-INFORMATION

Collected samples of material are no longer held in perpetuity by the company, organization, or institution that first acquired them but are increasingly being circulated through transfer or rental to third or fourth parties who may have an interest in the commercial exploitation of such materials. Scientific and academic institutions have a long history of collecting specimens and exchanging them with other, similar institutions as part of a broader project for scientific advancement. The protocols that have historically governed the exchange of these specimens, which include museum loan form agreements, and museum exchange agreements vary in complexity, but most simply constitute a record of specimens dispatched. Others are more detailed, specifying the length of time for which the specimen may be loaned, instructions for care of the specimens, requirements for shipping, and so forth. Few contain any stipulations about the use of materials, apart from the request that the providing institute be acknowledged as the source in any relevant publications. The charges levied to supply the materials have historically been minimal and designed to cover only handling and transportation expenses.

During the mid-1990s, these procedures and agreements began to undergo a radical transformation. The largest museums of natural history, those that are most aware of the implications of the trend toward the "remining" of scientific collections, began to develop more sophisticated policies governing the use of loaned specimens. In 1995, for example, the San Diego Museum of Natural History, introduced a new forty-page policy relating to the use of materials in its collection. It established strict terms governing access to and any potential sampling of material held in its collection and introduced specific clauses expressly prohibiting illicit trade in collected materials. Despite these refinements, the vast majority of collecting institutions do not yet have effective means of monitoring how lent materials are used in either the short or long term. Although scientific specimens are usually exchanged only for taxonomic study, some institutions do allow specimens to be examined for "research purposes," and they acknowledge that this may include sampling or dissection of the specimen. In addition to this, as the evidence produced by the ASC survey suggests, some institutions are now allowing corporate interests to access their collections, either directly or indirectly.

During the mid-1990s, institutions such as InBio became so concerned about the unauthorized circulation of collected samples to third and fourth parties and the loss of benefits associated with this practice that they began to require their institutional partners in the United States to sign an agreement prohibiting them from engaging in this activity.[54] The NCI also became uneasy about this issue. Believing the repository at Frederick to be a resource of the greatest possible value, they had actively encouraged industrial users to access it in order to obtain "limited quantities" of archived material for screening and testing purposes. As Neil Belson remarked earlier, judging just how little material to give a third-party recipient without risking their "taking it and running with it and never needing us again" is a delicate matter. The NCI became aware that third-party recipients might engage in "unlicensed copying" of their materials and consequently began to develop a separate contractual agreement that would regulate the terms under which collected materials could be circulated to other parties. This new "Materials Transfer Agreement" obliges any recipient of the material to agree to return "royalties and other forms of compensation" to the country that supplied the original sample of material to the NCI, "in an amount to be negotiated with the NCI in consultation with the host country organization."[55] They are obliged to do so even in circumstances where products are developed through the replication or synthesis of the original material.[56]

In reality however, the NCI has largely abdicated responsibility for monitoring compliance to the agreement or helping source countries negotiate appropriate terms of compensation with the new recipients of the material. Kate Duffy Mazan explained:

> You know it really is between them [the third or fourth party and the source country] to determine that. . . . The thing is we don't actually know what the terms of the agreements are—it is only if someone has sent us a copy. We really try and stay out of it; we don't like it if companies are sending us copies of everything they are doing with the source countries for fear that it is going to look like we

are calling the shots because we don't want to be. But they might pay them much the same way we do—an initial up-front payment of X amount. Then when we sign a license, they get a further payment et cetera. I really can't say for sure.[57]

The original collectors of these materials for the NCI have expressed some disquiet about the NCI's decision to circulate them to other consumers and to absolve itself of responsibility for monitoring the terms under which those materials are subsequently used. They are concerned that material that they collected for the NCI, under a particular contract and for a specific recipient, is now being circulated to others and that source countries might only know of these successive uses if, or when, third parties choose to negotiate a compensatory agreement with them. Further exchanges of the material may also occur—from third to fourth parties—without the knowledge or direct involvement of the source country as long as they, too, are subject to further MTAs. Source countries must rely on those parties to given an honest assessment of the viability of any materials they have acquired and screened through the NCI repository and to honor the terms of the MTA. As one of the senior researchers who collected material for the NCI has noted, this is not entirely satisfactory for either intermediary institutions or source countries:

> Those "Material Transfer Agreements" mean that they [the NCI] are taking some of those samples that were collected and they're providing them to companies outside of the original agreements and that is a very serious issue for us. . . . But you know, we've given up control of them by giving them to the United States government who'll presumably follow a series of rules—so we have no control anymore, we have no control in the same way that the village that they were collected in has no control anymore. We've worked with them [the NCI] to put a series of mechanisms in place to make sure that maybe they won't be abused. . . . We hope—but you'd have to talk to them about that.[58]

It is extremely important to remember, however, that the NCI's requirements for accountability (the MTA agreements) are the most stringent available in this industry. Unlike the NCI, the vast majority of scientific collecting institutions does not have any effective means of monitoring how materials on loan are used in either the short or long term. In light of this, it has been suggested that they, too, should be obliged to implement MTAs whenever specimens are exchanged or given out in loans. This suggestion seems to have polarized opinion within the collecting community. For some, such as Francesca Grifo of the American Museum of Natural History, such a proposal seems excessively restrictive — although she is conscious of circumstances in which such a requirement could become a necessity:

> For purely scientific research, why make every scientist crazy trying to do an MTA? I mean, that to me is absurd. It constrains research, and there is always going to be sleaze balls out there, whether you have an MTA as a requirement or not; there are always going to be people who completely skirt it. I think to constrain scientific research with those kinds of agreements, as long as it's basic research, is a misuse of time and energy. *But*, there should be the implicit and explicit notion that if a collection *changes in its purpose* from a scientific to a commercial purpose, then you should then have to do an MTA.[59]

However, for others, such as Henry Shands at the USDA, the ramifications of unlicensed copying or sampling of material extracted from developing countries are so profound that they unquestionably warrant the introduction of stricter regulatory protocols:

> Say a scientist provides somebody else with a sample from their laboratory to use. As I said, in the past that wasn't a problem, but now it has become almost *imperative* that an MTA go with it that limits the right of the receiver to do things with it . . . because even if someone [an industrial user] takes it and does something with it . . . then that might then prevent them [the source country] from doing something with it — if it's patented, for example. Then, of course, they [source countries] wouldn't have any rights to those components or to receive any royalties on them either. The CBD attempts to make people aware of all of that, but it's going to be a few years before we get a system in place *globally* that can cope with it.[60]

Although the largest collecting agencies, and those that are subject to heightened public scrutiny, have voluntarily introduced MTAs, there is, at present, no legal requirement for companies or institutions to use them or to otherwise monitor how materials that they exchange, or rent out, are subsequently used. Many small collecting agencies, brokerages, and rental firms profess to be supportive of the principle of benefit sharing; however, few have formal MTAs in place. This is possibly because some companies have no intention of going to the additional expense and effort of ensuring that the third and fourth parties that they sell or hire to then return compensation for the successive uses that they make of collected resources. This is, after all, another layer of bureaucracy that might deter prospective consumers from using their collections. They may also decide that they would prefer not to have to share the income stream. The fact that some companies and institutions engage in these dubious practices came as little surprise to Steven King:

> There are people out there who are trying to present themselves as highly reputable . . . but in fact they are turning around and selling those pure compounds without being up front about it to the collaborating institution in the source country and that includes [named institution], who are passing those results and compounds under contract to a pharmaceutical company—that is causing a lot of difficulties. But as far as regulating that goes, it is very difficult because it so hard to regulate anything that is so difficult to track. I mean, it has to be based on good faith to a certain degree. Anyway, that's the sort of thing that is causing a lot of difficulties in developing countries.[61]

Even when companies are well intentioned and introduce such agreements, the ability to extend production over time and space may present insurmountable barriers to their effective execution. The lead time for drug development is currently said to be ten to fifteen years. With the introduction of cryogenic storage, this lead time could be as long as 110–115 years or even longer. Can MTAs possibly remain effective over this period of time? Many working within the industry already harbor serious reservations about whether these agreements can withstand even the minor corporate restructurings that may occur over the next decade. Bob Borris from Merck observed:

> You can envisage a situation wherein small company X goes out and collects materials with all the appropriate material transfer agreements, royalty agreements, et cetera and then sells samples to major company Y and then basically consumes all of their capital and goes out of business. Big company Y would then have a situation wherein the company that they bought the materials from no longer exists, so the contracts that they signed really are not en-

forceable any more, so that could make things a little bit on the sticky side.[62]

While it is possible that large organizations such as the NCI will still exist in several hundred years' time and will, in addition, have the resources and the resolve to track how collected materials are used to ensure that relevant compensation agreements are honored, smaller companies are not always in a position to make such commitments. Steven King has admitted:

> If we go out of business totally, now or in the future, there is a certain undeniable risk aspect to this for all parties concerned. If we go out of business, it means that the people that we entered into this agreement with—well maybe the collections will go to some big company and maybe nothing will come of it—maybe the materials will go somewhere and someone will discover a drug but maybe no one will ever know because they won't reveal it—there is that risk aspect.[63]

King's comments proved to be uncannily prescient. Shaman had been developing a number of leads for pharmaceutical development, one of the most promising of which was Provir, a treatment for diarrhea. Despite having successfully concluded advanced clinical trials of the drug in 1998, the Federal Drug Administration demanded that Shaman double the number of patients for third-phase trials. Advised by the FDA that they were still several years and some $10 million away from marketing the product, Shaman was forced to regroup. The company divested itself of its pharmaceutical operations, relaunching itself as Shaman Botanicals, in part because the standards for testing botanical medicines are much less stringent than are those for pharmaceuticals. As a consequence of this restructuring, Shaman Pharmaceuticals announced in 2001 that it would conduct a sealed-bid auction of the company's intellectual-property portfolio, which was described in the on-line catalogue as comprising rights to six treatments, all of which "have been developed from the work of its ethnobotanists and physicians doing field research in Africa, South East Asia and Latin America . . . from promising plants that already have a history of safe and effective use as medicines."[64] The terms of sale of the portfolio may or may not include requirements for benefit sharing; however, no such entailments are mentioned in the conditions as they are set out in the catalogue. Whether any purchaser of the rights will consider that they have either a moral or a legal obligation to fulfill commitments made by Shaman remains uncertain.[65] As one inter-

viewee concluded, once collected samples begin to circulate, there is simply no sure way of establishing their provenance or of ensuring that agreements will be honored, or royalties disbursed. As he put it, "if you are talking about physically proving, I don't see any way in the world that they can.
About the best that you can hope for is a relatively consistent paper trail,
and, beyond that, you basically have to have a small chunk of faith."[66]

CONCENTRATION AND CONTROL:
PATENTING COLLECTED MATERIALS

The existence of a contract always implies a reciprocal relationship. This is
as true of bioprospecting contracts as it is of any other type of commercial
contract. Companies or organizations offer source countries compensation
for the collection and use of their genetic and biochemical materials—but
what is it that these companies and institutions receive in return? In exchange, they secure a valuable right—the right to control access to material
or even information derived from collected biological materials at a time
when they are still comparatively "exotic" and therefore most valuable. All
of the compensatory agreements that have been developed to govern the use
of genetic and biochemical material in the U.S. pharmaceutical industry
contain a clause that entitles the collector to apply for patents on all inventions developed under the agreement (see, for example, clause 6 of the NCI
agreement). A popular argument offered in defense of collecting projects is
expressed by a director of one of America's leading botanical gardens: there
is nothing inequitable about collecting biological materials as "the resource
is still there for others to capitalize on."[67] In other words, he suggests that the
resources are not in any way monopolized or alienated—an argument that
is frequently supported by reference to the fact that naturally occurring organisms cannot be legally patented. However, such analyses do not offer an
accurate definition of *what* is subject to monopolization.

As I have argued here, what is of most value to collectors is not the whole
organism but rather the material or information that can be derived from
it—for example, biochemical compounds or information about the structure of those compounds. As Rosenthal recently noted, patents are thus most
frequently sought for "technological advances embodied in the isolation and
modification of useful chemical derivatives and analogues of compounds
originally isolated from a plant, animal or micro-organism, for specific, identified uses."[68] So while it is true to say that the whole organisms are not being monopolized, the far more valuable material and informational re-

sources embodied within them certainly are. Source countries are at liberty to use the whole organisms in whatever ways they choose; they can go on selling them in markets as sources of food, fiber, or fuel. Local communities can continue to utilize them in their everyday lives, and, as long as they don't seek to make any other uses of them, there should be no conflict of interest. However, should they acquire the technology and capability to act on those organisms in more sophisticated ways, for example, to derive or produce from them certain types of genetic or biochemical extracts or information, they may well discover that a foreign company or organization will already have laid claim to them.

It is often stated, and indeed assumed, that the award of a patent is an appropriate compensation for the considerable investment of intellectual energy and creativity necessary to create a new pharmaceutical. It is evident, however, that patents are frequently awarded for compositions of matter—such as biochemical derivatives—in circumstances where the sole contribution of the inventor has been to isolate that compound from the organism in which it is embodied and to establish its novelty with existing systems of Euro-American property rights law. As I noted earlier, there are some interesting parallels here in the way that old and new centers of calculation exercise control over these collected resources. In a previous era, collected materials were "disciplined" by rescuing them from their chaotic natural surroundings and ordering them within a culturally determined system for codifying knowledge—the Linnaean system of classification. New centers of calculation now discipline biological materials in a different but equivalent way, by subsuming them within another culturally determined system for codifying knowledge—the Euro-American system of intellectual property rights law. Value accrues to those individuals who are first able to situate collected biological materials (or, more correctly, their genetic or biochemical material or bio-informational constituents) within this new system.

It is irrelevant if the materials or specific uses to which the materials can be put have already been codified in another culture or under another system for codifying knowledge. The reward—the right to exercise exclusive control over these particular materials and information—will only accrue to those who are able to establish the novelty of either *within this particular system for codifying knowledge*. Kate Duffy Mazan gave a very accurate illustration of how this process of alienation works within the bioprospecting industry:

> Under U.S. patent law, if we have a collector that goes out and talks to people in the source country—even if they are talking to a com-

munity or an indigenous group or whatever—who are saying we use this plant for this or that—if our collectors go and collect those and we grind them up and *we figure out what it is*—then the only inventor in that whole chain is our guys *who have determined what that compound is.* At that point, nobody has any interest in that patent except our guys, so there is no way for the suppliers to get any benefit back because they are not inventors on the patent so they won't get any royalties back when they license out the patent. That is why we stepped in with the letter of collection to say: "Look, when we license this out we know you can't be inventors under U.S. patent law, but when we license this out we will require that benefits be set." The only issue we've had is that in countries where they have signed the LOC we can give that guarantee, but in countries where, for whatever reasons—they're in a state of flux or they don't want to sign the LOC or whatever, we can't guarantee them that we are going to be able to force licensees to send benefits back.[69]

As Rosenthal correctly notes, it is impossible to obtain a patent on a "discovery per se" on "products of nature" or on other "physical phenomena."[70] In order to qualify for patent protection as a new composition of matter, the material must constitute an "invention" that is both novel, and capable of industrial application. However, as we have seen, the question of what constitutes "novelty" and what degree of intervention is necessary before a material can be determined to have been "invented" rather than "discovered" is a highly contentious one. As Mazan's comments confirm, "novel," in this instance, simply means previously uncharacterized within the American system of intellectual property rights law. In many instances, individuals or communities in source countries have already "figured out" (although not scientifically characterized) what medicinal uses particular plants have—which is exactly why they have been targeted for specific ethnobotanically based study and collection programs. Both Kate Duffy Mazan and Dianne De Furia, a senior executive at BMS, have confirmed that it is not necessary for collected material to be subject to any complex process of modification or "reinvention" in order to be successfully patented under a composition of matter claim. As De Furia confirmed, the material does not have to be modified in any way: "No, no, not at all—it can be just isolated. Your patent may even include a variety of derivatives of it—on the other hand, it could just be the parent compound." [71]

That so little invention is necessary to secure exclusive rights to the use of these compounds has been the source of some anxiety, not only for the countries and individuals who have supplied the compounds but also for those involved in collecting them, as this exchange with Gordon Cragg at the NCI revealed:

> **G.C.:** Well the thing is, if a compound is sent by the supplier and a test is done and some very interesting and novel result comes out of that, then the *NCI* could claim *sole* inventorship rights, even though the supplier provided the compound. This is following U.S. patent law, but a lot of people have a problem with that; they say "Well, my gosh, I supplied the compound, so why can't I be a co-inventor on that?"
>
> **B.P:** And why can't they?
>
> **G.C:** Because according to U.S. patent law, the invention claimed might be the particular interesting result observed from this compound, but the actual experiment in conducting that test was— well the only people involved in that particular experiment were NCI employers or contractors. The supplier provided the material but they were not involved in the actual invention.[72] I have to say I have a bit of a problem with that myself—but this is the way patent law works. A lot of suppliers from developing countries, I know, have a great deal of difficulty with the concept that they have supplied the material—and then admittedly NCI conducts the experiments—but still, they have no co-inventorship rights. There is just this feeling of unhappiness that they supplied a compound but that some other organization could claim sole inventorship for something done with that compound, and I can sort of understand that feeling. I don't know. . . . This definitely has been a sticking point in negotiations with organizations in supplying countries, particularly developing countries. They worry about whether this is another attempt on the part of developed countries to take something away from them.[73]

Consequently, most of the patents granted for biochemical compounds and gene fragments isolated from materials collected under contemporary bioprospecting agreements have been granted not to organizations, individuals, or institutions in source countries but rather to the large corporations, academic and scientific research institutes, or government agencies located in the United States that instituted the collecting programs. These new centers of calculation continue to play an important role as gatekeepers, although they now act to regulate the flow of valuable bio-informational resources rather than the movement of whole organisms. Interestingly, while in the colonial era power and profit could accrue from an ability to rapidly circulate whole plants or animals to various strategic locations, power and profit may now derive from an ability to actively restrict the circulation of extracted bio-information. In this era, competitive advantage in the pharmaceutical industry does not always come from being able to rapidly recirculate bio-informational resources, although, as we have seen in the case of the rental of such material, this can sometimes be so. In many instances, however, it is more useful to be able to control access to this information on terms that are advantageous to the collector, which, as this pharmaceutical company executive explains, is exactly the right that a patent affords: "A patent is a legal document that doesn't give you the right to do anything— it gives you the right to exclude others. So if I hold a patent and there are no licenses to it, I can exclude everybody else from using this unique compound." [74] Only when favorable terms and conditions for the licensing of the material have been set by the patent holder will the material or information be released for applied use.

THE COMPLEXITIES OF "CO-INVENTORSHIP"

Although it remains unusual for indigenous groups or collecting agencies in source countries to be named as co-inventors on patents, the NCI has stated that they will attempt to include them as such, where appropriate.[75] A right to co-inventorship is usually only granted in instances where the supplier is deemed to have contributed not only material but also detailed knowledge of the specific uses to which the collected material may be put. Conservation International, for example, when announcing their involvement as a partner in the new ICBG agreements in 1993, declared that under their benefit-sharing agreement, local shamans would be eligible for "patent and joint patent rights."[76] The sharing of what would otherwise be an exclusive property right with such suppliers has been characterized as a generous act—one that might afford suppliers a greater degree of control over the use of their knowledge and resources. However, evidence from this research suggests that this proposal remains, in most instances, hopeful rather than practicable.

As Dianne De Furia of BMS has suggested, companies are reticent to share rights with suppliers unless there is some objective method of validating their claim to have made some intellectual contribution to the development of the drug.[77] Second, there is the difficulty of identifying which particular individual or individuals might warrant inclusion as a co-inventor. As the system of intellectual property rights law is based on the principle of recognition of an individual inventor, complexities inevitably arise when attempts are made to include a large number of people, a community, for example, on an application. As a consequence, a disturbing degree of paternalism has crept into negotiations for the granting of co-inventorship rights. Evidence suggests that in several instances, the right of co-inventorship has actually been granted to the *ethnobotanist* responsible for collecting the material and information and not to the supplier of that material. Gordon Cragg of the NCI outlined two examples of this practice:

> We have had instances of that—do you know [named ethnobotanist X]? [Yes, I do] And you know he's been doing that Samoan plant collection program? Well there was this particular molecule, prostratin, which came from a plant used in Western Samoa by their healers for treatment of what they call yellow fever, but we think actually that it's for something that is more like hepatitis, but, nevertheless, the plant was selected on the basis of that knowledge. Now in the patent on prostratin, interestingly enough [X] is a co-inventor on that, not the actual healer. Now I don't know what the actual reasons for that are, and I would trust [X] implicitly, that he has the interests of all the indigenous peoples—the community that he works with there, their interests—very much at heart, but the actual healer wasn't the co-inventor, [X] was, because he supplied us with this particular ethnobotanical knowledge.[78]

Commenting later, on the way in which benefits would be disbursed under the new ICBG agreements, Cragg again confirmed that an appropriate distinction is not always maintained between the individual who provided the material and/or information, and the individuals or organizations that purport to represent their interests:

> **G.C.:** My guess was that if a plant . . . collected on an ethnobotanical basis did yield a compound of interest then someone, whether it was going to be the person that originally divulged the knowledge—the healer, or say someone from Conservation International—I know [named person] isn't involved with that any longer but *whomever*—they would have a distinct claim to co-inventorship.

B.P: Who, Conservation International?

G.C.: Well yes CI, but it would have to be some individual—whoever that individual was from CI. This whole idea of including healers as co-inventors is tremendously difficult. I mean what if it's community knowledge? This may be why [X] claimed co-inventorship, rather than some Western Samoan person—perhaps because it was community knowledge. But with someone like [X], it is good to have him as a co-inventor because then if anything does come of this compound there is no argument that he and his university have rights there, and I know that the university has, well [X] actually, has ensured that there are agreements between the university and the communities such that if anything comes of this the communities are definitely going to get their fair share."[79]

Although in this instance there seems to be no reason to be unduly concerned about the motive of the intermediary, X, there can surely be no certainty that every person who presents themselves as a legitimate representative of an indigenous community will not be seeking to capitalize on their ability to insinuate themselves into the patent process. It does seem remarkable that they should be able to do so in such a way that it is *they* who ultimately hold a right of co-inventorship and not the indigenous person or people whose knowledge first contributed to the identification and commercial development of the drug or therapy.

Perhaps the greatest impediment to the use of co-inventorship or sole inventorship rights as a mechanism to secure indigenous rights of control over these resources is the inability of indigenous communities to effectively defend those rights. As several interviewees have pointed out, while the grant of co-inventorship rights would *theoretically* enable individuals from source countries to exercise joint or exclusive control over the uses that will be made of collected materials, there is a significant difference between the ability to *hold* a right and the ability to *exercise and protect* that right. Although it may appear to be paternalistic to deprive suppliers of the opportunity to apply for co-inventorship or sole inventorship status in their own right, the decision to do so can be well intentioned and made out of a concern to ensure that the patents granted are ultimately enforceable. As Gordon Cragg remarked, "a lot of people don't appreciate this, especially in source countries, but just to apply for the patent and then to defend it if it's challenged is an immensely expensive project which most countries just can't afford. So even though the idea of having sole inventorship might seem very attractive because they have total control—to actually go though the whole process is not a trivial matter for most countries."[80] However, an inevitable consequence of this is that control of these resources will remain concentrated in the hands of large corporations, for, as Joshua Rosenthal suggests, it is only these large companies and organizations that have the interest, experience, resources, and expertise to successfully defend patent rights:

> The thing to remember is that patents are actually very expensive to file and maintain—at least 5,000 dollars a year—and then you have to have the capacity to defend it and the resources to do so, and so people who understand that are generally in agreement that they want to have that under well-funded, sophisticated, private companies. People who don't know a lot about bioprospecting— who are excited about it but nervous about what it means to their control of the resources—tend to want to have control of the patent, even if they don't have the ability to defend them.[81]

One last matter that is often overlooked in discussions of the grant of patent rights, and in the structuring and operation of benefit-sharing agreements more generally, is the question of expiration of the patent. Although a patent is an instrument that gives exclusive monopoly rights to a particular concept, process, or composition of matter, these rights are time limited. In the case of a patent granted on a composition of matter or product developed for use in the U.S. pharmaceutical industry, the patent would remain effective for a period of seventeen years. Almost all of the provisions for the payment of long-term compensation under existing benefit-sharing agreements are based on securing revenues from the sale of patented products. This raises the question of what will happen to the revenue stream once the patent expires. Even though Steven King's company Shaman subscribes to the principle of royalty payments, he was also concerned to acknowledge the effect that the expiration of a patent will have on benefit sharing agreements: "The danger is that after a patent expires it becomes truly a generic and then anyone can extract it or get it or make it, and then they can undercut the price, and then it [benefit sharing] is no longer financially viable, and any commitments that you've made to communities or countries that it came from, you can no longer protect, because the material is then in the public domain."[82]

In this chapter I have investigated the way in which new contracts and compensatory measures governing the use of collected genetic resources have been generated and implemented and provided an assessment of their likely efficacy. While the biodiversity convention established general protocols relating to the way in which genetic resources might ideally be used, it provided no clear guidelines as to how these protocols could be translated into workable contractual and compensatory agreements. Evidence produced here shows that while countries are at liberty to develop and implement their own contractual and compensatory agreements, most have been encouraged to adopt the three-phase model of compensatory agreements developed by U.S.-based institutions such as the NCI. A closer examination of the history of the development of such agreements revealed that they have been devised in the United States by a small group of elite actors within the bioprospecting industry, many of whom hold senior positions within what I have termed "the new centers of calculation." It would seem that the institutions that are centrally involved in the collection of genetic and biochemical material materials have managed, through these actors, to acquire de facto responsibility for determining the terms and conditions that now govern this new trade in bio-information. As a consequence, it is perhaps inevitable that the agreements primarily favor the interests of these collecting institutions and organizations. These agreements have rapidly become hegemonic over both space and time, coming to constitute what Featherstone would term a "proto-universal culture of regulation." The successful universalizing of this system for "disciplining" biological materials has been effected, as it was in an earlier era, through the vector of a linked network of collecting agents and institutions. These agents and institutions have played a central role in promoting, mobilizing, and disseminating this particular form of regulation, introducing it to an ever-wider range of countries and cultures. Although these three-phase compensatory models have quickly become normative, it is by no means clear that they will provide an adequate mechanism for returning to source countries a "just and equitable" share of the benefits that have accrued from the exploitation of their resources.

There are, as I have set out to illustrate in this work, a number of reasons why these mechanisms are unlikely to be successful. As we have seen, current benefit-sharing agreements are based on a number of premises. Advance payments, which may prove to be the only form of monetary compensation actually received by the supplying community, are comparatively small. The justification for keeping them so is based in part on a presumption that collectors will return for more supplies of material once commercial production of the drug commences. As the evidence produced in both chapters 6 and 7 suggests, however, this probability is inexorably diminishing as new technologies provide alternative sources of supply. The compensations offered in the second phase—infrastructural support and training—are limited and can be construed as mechanisms primarily designed to expedite the collection and extraction of the desired genetic and biochemical materials. Longer term forms of compensation—such as royalties—are relied upon to produce continued returns for the successive uses that are made of collected resources. Unfortunately, however, evidence from this research suggests that being able to effectively track or monitor these successive uses will prove an extremely complex task, complicated by the fact that such materials can now be transported or circulated in a virtual state, stored cryogenically over long periods of time, replicated relatively unproblematically, combined or manipulated to form new versions, and sampled to form the basis of other products.

Collecting institutions are able to capitalize on the bio-informational resources embodied within collected materials in two important ways. First, they may restrict access to these resources by subsuming them within a particular culturally defined system for codifying knowledge that affords them exclusive and private rights of ownership over them. Having secured control of these bio-informational resources, they are then at liberty to recirculate and redeploy them repeatedly to further financial advantage. Unless existing regulatory agreements are reformed to include effective measures for tracking and compensating for the successive uses that these collectors make of bio-informational resources, they are unlikely to ameliorate the inequities that have traditionally characterized the exploitation of natural materials. Where then does this leave us, and where to from here?

7. BACK TO THE FUTURE

As I sit down to conclude this work, I have before me copies of the daily papers, each of which makes reference to the forthcoming Earth Summit on Sustainable Development that is about to commence in Johannesburg, South Africa. As exactly ten years have now elapsed since the Convention on Biological Diversity was first implemented, it seems an appropriate, and timely moment to reflect on how the introduction of this new regulatory regime has affected approaches to the utilization and stewardship of what has become, in the course of that same decade, one of the world's most valuable commodities: genetic and biochemical material and information. Interest in the collection and use of these resources began to escalate dramatically during the early 1990s as new technologies, new global environmental and economic policies, and restructurings in the corporate and scientific realms began to create new incentives to collect genetic and biochemical materials and new avenues for their exchange. The history of the collection and exchange of biological materials is extraordinarily long, stretching back hundreds of thousands of years. Within this context, it was easy to assume—and many have—that bioprospecting is simply a new descriptor for an age-old practice.

In this work I have tried to show that there are difficulties with this assumption. Although the collection and exchange of biological materials does indeed have a very long history, the application of sophisticated, indeed, entirely unprecedented biotechnologies has fundamentally altered the ways in which these resources are now collected and used—particularly within life-science industries such as the U.S. pharmaceutical industry. The change that has occurred in the way these materials are collected reflects a broader transformation that has occurred in drug development in recent years. The progressive introduction, and later dominance, of molecular approaches to the creation of pharmaceuticals and therapies has directed attention toward the components of living organisms. The genetic and bio-chemical materials and information embodied within them have proven to be extremely useful and valuable commodities, for both scientists and researchers, and interest has inevitably coalesced around the question of how to more effectively and efficiently collect and utilize these resources.

It soon became apparent to collectors that one means of improving the efficiency of the collecting process was to concentrate on collecting only those "key" or "essential" elements or components of biological organisms that were essential to processes of research and development, while divesting all other material that was considered to be extraneous or inessential. One way of achieving this was to produce a series of new engineered arti-facts—such as cryogenically stored biochemical extracts and samples of tissue, cell lines, and sequence information—that could act as proxies for whole organisms. Of course, the genetic and biochemical material and information embedded in whole organisms has always been available for use; it has just not been as accessible as it is when embodied in these new forms. I have argued in this book that changing the way in which this genetic and biochemical material and information—what I have termed "bio-information"—is rendered, or presented, has had a profound effect on the practice and politics of biological resource exploitation.

In thinking about the way in which biotechnologies have transformed trade in biological materials, I have drawn a parallel between the actions of biotechnologies and informational technologies. Both, I have argued here, are employed to act on complex phenomena in order to produce from them proxies that provide consumers with, if nothing more, then at least those "key" or "essential" elements that they most desire. In both cases, producing these proxies involves a process of distillation—a de- and rematerialization. Much of the existing material or body of the original is divested, enabling the remaining information to be rendered in new, more lightweight, mobile, and transmissible forms. Just as new informational technologies have enabled particular resources to be structured in ways that have made them easier to circulate, store, and reprocess, so biotechnology has also enabled whole organisms to be rendered in ways that make the genetic and biochemical materials and information embodied within them much easier to transmit, store, reprocess, and recirculate.

These processes are of the utmost significance, as they allow the collectors of biological material to speed up the social and spatial dynamics of collecting from which power and profit derive. As I began this work by arguing, collecting is an inherently political activity, in that it involves annexing particular materials for exclusive use. The collector's power derives from his or her ability to acquire materials of interest, to concentrate them within particular locations where they can be ordered, controlled, and disciplined, and to then recirculate them (or not) within the marketplace to strategic advantage. Any factors that allow them to speed up the different phases of this process will be all to their advantage. I have argued that the ability to create new, highly transmissible bio-informational proxies that, unlike many of their historical counterparts, actually prove fungible for them in processes of research and development has allowed a select group of collectors to acquire, concentrate, and recirculate these commodities with far greater ease. This newfound ability, as well as the creation of a new regulatory paradigm that creates conditions within which these novel commodities may be exchanged on normative terms, have together enabled collectors to create a burgeoning and lucrative market economy for these bio-informational commodities.

Almost as soon as it became apparent that the introduction of new biotechnologies would inevitably create a demand and a market for genetic and biochemical material and information, concerns were raised as to how this market would be regulated and in whose interests it would operate. Much of the collecting that took place in earlier eras was undertaken in tropical developing countries, and these countries have remained the target of collecting expeditions. Many of these states were anxious to avoid being the object of a new wave of biocolonialism, and they began to lobby intensively for the creation of new policies and regulations that might govern the collection and use of genetic and biochemical resources, hoping to ensure that these activities were carried out in a more just and equitable fashion. This pressure was, in time, translated into the regulatory protocols that were introduced under the biodiversity convention and that have created the paradigm within which approaches to the governance of these resources have been shaped. During the late 1980s and early 1990s, an elite group of senior practitioners within the bioprospecting industry began to translate these protocols into a series of novel contracts that established the terms and conditions under which these resources could be exploited. A principal feature of these agreements, and one that played a central role in legitimating this new era of bioprospecting, was the inclusion of a formal benefit-sharing regime.

Some of the most interesting and significant characteristics of this emerging trade in bio-informational resources are the modes of transaction to which it has given rise. Bio-informational proxies, as I have noted here, privilege the genetic and biochemical information embodied in whole organisms at the expense of other attributes of the organism. This information (whether in a partially or wholly decorporealized form) may be used *successively* by a variety of consumers for a variety of different purposes. Collectors have quickly recognized that they may exploit their assets without relinquishing complete control of them. Rather than collecting and selling samples of material outright, many prefer to create an ongoing revenue stream by repeatedly selling *access*, on a short-term basis, to the genetic or biochemical information contained within them. This has given rise to forms of commodity exchange that are unprecedented in this domain, such as rental, licensing and pay-per-view. As I have illustrated here through empirical example, these resources are particularly valuable, as they can be used to form the basis of new products and processes.

If this trade is to be an equitable one, these benefit-sharing agreements must provide a mechanism that enables supplying countries to share in the profits that are generated from the many successive uses that are made of the bio-informational resources that have been collected within their borders. Agreements began with an awareness of this—the inclusion of a royalty mechanism reflected an acknowledged need to meet this aim. However, as I noted earlier, it is impossible to pay a royalty unless it is possible to determine to whom that royalty should be paid. This necessitates establishing a chain of consumption: ascertaining with certainty from whom the material was collected, by whom it has been used, for what purpose, and with what outcomes. It quickly became evident that this would require the introduction of a separate set of formal mechanisms for tracking and monitoring all of these successive uses.

This was realized through the development and implementation of a further set of contractual agreements that have become known as "Material Transfer Agreements." The use of MTAs was initially confined solely to transactions of bio-information that were understood to be commercial in nature. Almost immediately, however, it became clear that the distinction between commercial and noncommercial uses of such material and information (if it ever had existed) was being dramatically eroded by changes in the operating environment within the life sciences, particularly by the creation of a host of new dependencies and collaborations that established new links across traditional boundaries (e.g., corporate sponsorship of academic research, shared commercial and academic use of resources, and so on). Pressure thus mounted to extend the application of MTAs to *all* transactions that involve the transfer of collected biological materials. Although both corporate and scientific researchers recognize why, if MTAs are to be effective, they have to be applied uniformly, I have also detected their grave concern that extending such mechanisms to every possible exchange of material will be immensely burdensome, potentially unworkable, and, as I shall go on to suggest, morally corrosive.

This "compensatory" model is such that it also creates a need to address questions of informed consent. It has been argued that it would be unethical to change the use of collected genetic materials (for example, from a scientific to a corporate use) without inquiring whether the original provider agreed to that change of use. Every such change may yield an economic return, and as suppliers are potential recipients of a percentage of that return, it seems important that they be fully apprised of and in agreement with any proposed change of use. Although I have not addressed the issue in detail in this work, the task of securing such consents from suppliers is an immensely complicated one. Even in the most simple scenario—where a supplier provides material or information for an immediate use—it must first be possible for the collector to identify the particular individual or individuals from whom consent must be gained. This entails addressing complex and perhaps irresolvable questions about who "owns" these bio-informational resources (for example, individuals, communities, or states).

Acquiring "informed" consent also necessarily involves being able to adequately describe to such groups (once identified) how the material will be used in ways that are intelligible to them, by which I mean understandable within the context of their worldview and life experience. These difficulties multiply, first, as the numbers of transactions increase over time and, second, as the array of potential uses to which the material could be put exceeds that which any one person might reasonably be expected to comprehend or assess. At present, collectors have two choices. The first is to secure a *prior* informed consent for all potential uses; however, these consents are likely to be so broad in their compass as to lose all relevance. Such agreements are, in effect, a permission to do all things with the material even if those things cannot be specified, and in that case, to what have suppliers consented and how could their consent be said to be informed? Alternatively, collectors can insist that each subsequent user of the collected material return to the suppliers (assuming, of course, the suppliers are still there to be asked) to secure their further consent, on each and every occasion that the material is used in a fashion not previously anticipated or agreed upon. This will necessarily involve providing adequate explanations of what are, potentially, ever more complex technological applications (for example, gene splicing, nanotechnology, and so on).

Policymakers have responded to these complexities by introducing more and more layers of procedure designed to accommodate and regulate every possible eventuality, including those that cannot yet be known. Nowhere is this more evident than in the newly negotiated "Bonn Guidelines on Access to Genetic Resources and Fair and Equitable Sharing of the Benefits Arising out of Their Utilization." The agreement, which represents the culmination of ten years of regulatory development within the biodiversity convention, is immensely detailed; the proposed protocols for governing access to and use of these genetic and biochemical resources now take some twenty-five pages to articulate. They include recommendations that collectors introduce, among many other things, legally consistent mechanisms for securing prior informed consents; systems for protecting and encouraging customary uses of biological resources in accordance with traditional practices; mechanisms for extending and strengthening the application of MTAs; frameworks for improving local awareness and capacity to implement benefit-sharing agreements; and, indeed, ever more complex proposals for benefit sharing, including specific provisions on type, timing, mechanism, and distribution.[1] Together they constitute a system that is, it could be argued, ever more convoluted, onerous, and unwieldy to implement, and that may, in fact, serve only to deter prospective consumers.

Despite larding up these layers of contractual and legislative requirements (royalty agreements, MTAs, informed-consent agreements) in what is an increasingly baroque regulatory framework, it remains the fact that supplying countries and communities have yet to receive any *substantial* economic returns from the exploitation of their collected materials. It could be argued that this is simply because no products have yet been developed from the material. This seems surprising, however, given that this new era of collecting and highly sophisticated investigation and exploitation of collected materials has now been underway for nearly twenty years. It is an established fact that in recent decades some 25 percent of all prescription drugs have been derived from natural products, most of which have been collected in foreign countries. Are we now being asked to believe that despite the introduction of much more sophisticated technologies, no new drugs have been developed from such materials over the last twenty years, or at least not in the last fifteen years, since access- and benefit-sharing regimes were introduced? An alternate explanation would be that these collected biological materials have, of course, formed the basis of many commercial products during that period; it is just that this has not been established factually or acknowledged. What this suggests is that the existing regulatory system has failed to play its redistributive role, despite its elaborateness.

As I have argued here, this failure did not occur because such agreements have not been well intentioned but rather because changes in the way biological materials are rendered and utilized have combined to make the task of monitoring and compensating for their use extremely difficult. This problem is not, of course, confined to the biotechnology industry. As I have shown, the introduction of other new technologies—such as informational technologies—has created remarkably similar difficulties in other industries. The ability to transfer books, music, and films into a digital form has enabled these materials to be circulated, copied, and modified with extraordinary ease. The Associated Press reported on 21 August 2002 that it was estimated that 10 million people had tried to download the new and aptly named installment of the *Star Wars* saga, *Episode II: Attack of the Clones*, in the weekend after its release using Internet relay chat file-sharing systems, and that 4 million of them had succeeded.[2] Media executives and attorneys are so determined to stamp out these practices that they are now devoting a considerable percentage of their research and development budgets to devising new methods to prevent unlicensed replication of these products, such as encryption or hardware modification. Despite this, the industry is still finding the problem to be largely intractable. Immense amounts of time, energy, and resources are being dedicated to the task with little apparent success.

If those working in the bioprospecting industry hope to be able to trace all the uses that are made of collected genetic and biochemical materials and information over both space and time, it would seem that they too will need to employ some equally advanced technologies. Ever more elaborate and sophisticated mechanisms for monitoring and tagging the movement of genetic material and information are certainly being developed. DNA encryption, genetic fingerprinting, electronic identification (EID), molecular markers, and DNA profiling are all being promoted as the latest and most effective tools for establishing and maintaining the provenance of genetic components or sequence information as they are passed from one prospective user to another. All of which raises a question: How far down the road of tracing the successive uses of bio-information could, or should, we attempt to go?

There is no question that genetic and biochemical materials are used extensively in the biotechnology and life-sciences industries; however, they are primarily used when in these partially decorporealized or wholly informational forms. If benefit-sharing agreements are to work, they must be able to trace and take account of all the transactions that involve the use of genetic and biochemical resources, even when they are rendered in these elusive forms. The new tracing technologies that I outlined above could be applied to this task (they are not currently), although this would be an immensely complex, expensive, and time-consuming undertaking. Even systems such as the sustainable-timber certification scheme developed by the Forestry Stewardship Council, and the livestock passport scheme introduced to trace the movement of large whole organisms (such as trees and cattle) as they are transferred from one location to another through a succession of market transactions, have been derided for regularly failing to establish an identifiable "chain of custody."[3] It is important to note that these projects have faltered even though the material that they seek to trace is infinitely larger and more stable than anything that one might attempt to trace in the bioprospecting industry.

Despite this, a large proportion of the (relatively) small amount of money that has been generated through bioprospecting ventures has been channeled into creating a flourishing bureaucracy devoted to the task of attempting to track, monitor, and secure compensation for the use of tiny fragments of genetic and biochemical material and forms of extracted bioinformation to equally questionable effect. Many of the NGOs whose principal aim is the protection of biodiversity, and many institutions in developing countries, are now devoting a considerable proportion of their operating budgets to the task of developing, implementing, and overseeing what may well prove to be an ineffectual set of procedures. Developing-state governments are not immune from this process either. Many now feel compelled to develop detailed legislative and administrative structures that, they are told, will enable them to secure ongoing returns from bioprospecting operations. Given that this regulatory paradigm may fail to deliver major returns and that the opportunity costs of directing finite resources away from other crucially important areas, such as education and health provision, are so considerable, it becomes even more important to undertake an unflinching assessment of whether such investments can be justified or sustained.

Recognizing the limits of the existing regulatory paradigm is clearly of profound importance to all concerned. However, these matters also raise a larger question: Is there a risk that the existing regulatory system may not only fail to meet its goals but may, in fact, exacerbate the very problem of mistrust that it sought to remedy? As I have illustrated here, the new era of collecting that began in the 1980s emerged in a climate of uncertainty—

there were hopes that it might provide new opportunities and returns for developing countries but also concerns that the projects might reproduce the colonialist practices of an earlier era. The new regulatory protocols that were introduced under the biodiversity convention were explicitly designed to improve relations of trust—they sought to remedy past inequities by ushering in a new regime of just distribution of gains. This original impulse has since been translated into layer after layer of ever more detailed and complex forms of contractual regulation and practical requirements for transparency and accountability that have become sedimented over the past decade into a set of practices that have now been adopted globally.

The task of conforming to these new protocols is onerous and immensely consumptive of time, energy, and resources. Many scientific research institutes and, perhaps more importantly, many companies are so constrained or wearied by the need to meet these requirements that they have elected to either abandon in situ collecting altogether or to engage in these activities illicitly. Neither course of action brings any financial benefit to source countries. Those companies, organizations, and institutions that do decide to comply are burdened by the weight of administering contacts that attempt to pin down, through ever more finely calibrated mechanisms, the whereabouts and recent uses of every last bit and byte of genetic material and information, when it must be clear to most that this is a truly Sisyphean enterprise. The structural inability of this system to produce the promised returns, will, in time, I believe, give rise to a serious, profoundly corrosive, and depressing sense of frustration in all parties.

The notion that it is possible to secure accountability and trustworthiness through the imposition of more complex systems of behavioral auditing is certainly not confined to this realm—it has been widely embraced in many areas of medicine, science, academia, and public life in recent years. As the moral philosopher Baroness Onora O'Neill suggests, "a prominent feature of this widespread movement to improve accountability has been an increasing reliance on more formal procedures, including contracts, letters of agreement and financial memoranda that impose highly complex conditions."[4] Formalization of procedure appears to have a number of advantages that are, as she suggests, "constantly mentioned by its advocates: mutual clarity of expectations, clear performance targets, defined benchmarks of achievement, enhanced accountability."[5]

What her insightful analysis highlights, however, is that the *failure* of these formal mechanisms to deliver desired outcomes may ultimately engender, rather than remedy, feelings of mistrust. Ironically, although a strict auditing of these transactions promises to deliver accountability through transparency, in fact what it implies is that those involved in such exchanges are somehow inherently untrustworthy and must thus be subjected to continuous external monitoring. Although this may rankle, subjects might be prepared to tolerate it if they believe that the application of more detailed forms of auditing and monitoring will actually deliver equitable working relationships, standards, and outcomes. As Thompson notes, however, most eventually come to believe that such procedures in fact "only create further levels of bureaucracy and inefficiency . . . and set in motion a process that may exacerbate rather than alleviate the problems they were intended to address."[6] This may, as he concludes, contribute to deepening a culture of distrust.

This argument can, it would seem, be applied to an analysis of the development and application of regulatory paradigms in the bioprospecting industry over the past decade. These were introduced primarily to ameliorate concerns that source countries had about the inherent untrustworthiness of collectors—an attitude that had its genesis in the colonial era. In the interests of ensuring that the contemporary bioprospecting operations were carried out in a just and equitable fashion, stakeholders began to devise mechanisms that would regulate access and ensure an equitable distribution of benefits. These mechanisms and attendant procedures have become progressively more formal and more elaborate as the years have gone by, and this has, perhaps, ironically engendered further feelings of mistrust. These feelings are at risk of being exacerbated by that fact that *despite* the introduction of these increasingly complex procedures, remarkably little money has been returned to source countries from these bioprospecting operations.

Although we continue to receive estimates of the truly astonishing amounts of income that have been reaped from the sale of products developed from natural materials sourced, in most cases, from developing countries (estimated to be between $75 and $150 billion annually in the pharmaceutical sector alone), it remains the fact that none of these countries have received any significant proportion of this income—these increasingly convoluted regulatory and compensatory mechanisms have largely failed to produce the desired outcome.[7] An unfortunate consequence of this is that developing countries are being affirmed in their belief that this new wave of bioprospecting is unlikely to return the benefits it promised. Their confidence in the project has been so undermined that many are now threatening to close their borders to all collectors, including those that are undertaking collections for scientific purposes such as taxonomic identification. It would be truly unfortunate if this were to occur, as such activities are central to the project of preserving global biodiversity.

There is also clear evidence that overregulation is deterring companies from pursuing natural-products research. Rather than opening up new possibilities for the creative and sustainable use of biological materials and information, these new multitiered levels of regulation act to progressively choke off such opportunities. There may be those who are of the view that this is no bad thing—that biodiversity is best left unexploited. However, in an increasingly globalized world, the likelihood of biodiversity remaining so is remarkably low; it would be naïve to think otherwise. The question, it seems, is not whether these resources will be exploited but rather how they will be exploited and how this market might be organized such that it operates on a more equitable basis. Given all that we have ascertained from this research, it does not seem inappropriate to propose a radical rethinking of approaches to the regulation of this burgeoning global trade in genetic and biochemical material and information. I believe that by going back to the past, it may be possible to see a new way forward into the future.

Prior to the introduction of the biodiversity convention, access to genetic and biochemical materials was unrestricted. Countries, collectors, institutions, and individuals all held an equal right to access these resources, investigate them, and employ them in any number of creative ways. There seems to be a consensus of opinion that this was generally desirable—that the ability to access them freely stimulated innovative and inventive approaches to their use. It was also evident, however, that the power to utilize these resources had historically, and would in the immediate future, continue to be concentrated in the hands of those collectors with links to the most sophisticated laboratories in the developed world. This inequity created unequal power relations and a genuine probability that most of the profit that derived from the exploitation of these materials would remain in the hands of those individuals. A property-rights war broke out as a consequence: supplying countries began to claim rights over their genetic and biochemical resources, while corporate interests began to claim rights to the engineered artifacts that they created from them.[8] The outcome was the introduction of new benefit-sharing regimes designed to ensure that a proportion of the profits that accrued from the commercial exploitation of these collected materials be returned to those countries that had supplied them.

The principle of sharing benefits with supplying countries is not, I believe, disputed by any party, even those companies that would be expected to relinquish a portion of their profits to meet this requirement. In the course of my research, I did not find any executives who were opposed to paying a royalty of between 1 percent and 5 percent of net profits on products derived from natural materials that were not collected domestically. Most large pharmaceutical companies (those that generate large numbers of products and profits) now fully accept that they must pay for their raw materials. Ideally, they would prefer to only pay for those that prove to have some long-term utility for them, and this is understandable. They are unconcerned about the addition of this extra cost, in part because it is unlikely to be borne by them directly—it will simply be added to the sale price of the product as a production overhead, alongside costs such as advertising and packaging.

Companies are also now acutely aware of the adverse publicity that attaches to attempts to secure raw materials without paying some appropriate form of recompense to the supplier, and they have no desire to attract such opprobrium. Most are concerned to act as responsible and environmentally conscious corporate citizens. What companies emphatically *do not* wish to do, however, is to spend ever greater proportions of their operating budgets complying with an unnecessarily cumbersome and unpredictable regulatory system. Regulatory protocols (such as benefit-sharing agreements and MTAs) are being continually reworked and refined to take account of new developments. Understanding and implementing these changes is an enterprise that absorbs an increasing number of administrative and legal personnel in corporate, scientific, and academic organizations, whose services may be more usefully deployed on other tasks.

Source countries find themselves in a similar position. What they ideally want is to receive some significant and ongoing economic returns for the use of their resources, without having to commit an inordinate percentage of their extremely scarce resources to the task of developing, administering, and monitoring new policies and regulatory regimes. Although I have not had an opportunity to calculate it here, it would be interesting to conduct an assessment of how the costs that supplying countries have incurred in setting up and implementing bioprospecting legislation, policies, and agreements compare with the returns that have so far been realized from such ventures. Such an analysis would almost certainly reveal that the system is in negative equity. If the system as it currently exists does not appear to be satisfactorily meeting its aims, then other, alternative schemes must be proposed.

What I offer here is nothing more than one possible alternative derived from this detailed critique of current practice. Although radical in conception, it retains and respects the original impulses that informed the development of new regulatory approaches to the collection and use of genetic and biochemical resources, such as access and benefit-sharing agreements. I offer it not as a fully realized proposal, for it is clearly not, but rather in the hope that it might stimulate some further debate and discussion about the shape that future regulatory paradigms in this field might take. The proposal that I intend to outline here is informed by ideas being developed in parallel domains such as the media and software industries. They have been devised in response to the need to evolve more sophisticated ways of dealing with the use, both licensed and unlicensed, of commodities and resources that have proven to be highly transmissible, readily replicable, and, as a consequence, extraordinarily slippery.

It seems to me that what matters most to collectors and suppliers in the final analysis is the question of what products are developed from collected natural materials. Samples of material and information may be used in many speculative endeavors; they may prove to be of use in some and not in others. All that really concerns people is whether those materials go on to form part of a patented or copyrighted product—such as a drug or sequence database—that is marketable and that generates profits. Suppliers have a right to expect that in such instances they would receive a small royalty for having provided the resources on which these products are based. Within the bioprospecting industry, royalties range between 1 percent for material collected at random to 15 percent for material that already has indications of proven efficacy against specific diseases. However, there is a general agreement, and evidence and probability would seem to suggest, that only a comparatively small proportion of collected samples would fall into this latter category. The majority of collected samples currently attract royalty payments of between 3 and 5 percent.

I propose that we abandon the task of attempting the trace all of the myriad uses that are made of collected genetic and biochemical materials and information and concentrate instead on working to secure a voluntary, global agreement from the pharmaceutical industry that they will add a sum of between 3 and 5 percent of their profit ratio to *all those products that they currently have in the marketplace that are based on collected natural materials.* This levy would remain in place for as long as those products are sold. In order to ameliorate some of the inequities that have characterized collection processes in the recent past, it does not seem inappropriate to suggest that this levy also apply to products, such as vincristine and vinblastine, that were derived from materials collected without recompense but which continue to generate substantial profits for drug companies.

I would suggest that a similar duty be levied on users of genetic-sequence databases and indigenous-knowledge databases. This could be easily added to the charges that institutions should, and do, routinely require to access these types of information. These levies, which would amount to not more than a few cents on most proprietary products or informational transactions, would be paid directly into a superfund that would ideally be administered by a global regulatory agency such as the Global Environment Facility.[9] Countries and communities from the developing world might then apply directly to the fund with proposals for development and conservation projects.

As I have suggested here, the task of attempting to establish the provenance of collected materials is highly problematic, and it would not be fruitful, in my view, to devote too much attention to trying to establish which countries had provided what material or information. There will undoubtedly be cries of indignation about the idea of abandoning a strict auditing of who gave what to whom and from where. Some countries will argue that they have given much more material than others, while others might argue that, while they have given less, the material that they have given has proven to be much more valuable. While such claims may well be true, it will be impossible to determine their veracity unless we are able to establish, with complete certainty, how those materials have in fact been used over time and space—a task that requires access to actuarial information that will always remain just out of reach. While we have awaited the successful completion of this potentially unworkable auditing process, years and years have slipped by without the production of any form of substantial compensation.

This scheme, might, I believe, yield a number of substantial and immediate returns for both suppliers and collectors. It is my fervent belief that what supplying countries most desire is to secure access to an ongoing income stream that they may utilize for a variety of conservation and development projects that can provide ecological, technological, cultural, economic, or social benefits. We all, collectively, live on a small and increasingly imperiled planet. As processes of globalization advance, we all become much more dependent on one another—environmental and economic collapses in one region of the world can have catastrophic consequences in another. There would undoubtedly be complexities associated with the disbursement of these funds, and I do not wish to downplay them here. The world's more biodiverse countries or those that have played host to a greater number of contemporary collecting projects may wish to argue for a greater proportion of the funds, and they may be justified in doing so. My abiding sentiment, however, is that most citizens of the world would feel that a considerable gain had been made if income from the superfund were used to reverse imminent environmental destruction, *wherever that may be occurring*, and I feel certain that consumers of pharmaceuticals would consider this an extremely appropriate and welcome use of any small surcharge that had been levied upon them.

Another considerable advantage of this scheme is that it would yield an immediate, substantial, and ongoing revenue stream. Instead of having to wait twenty years or more for income that may never be realized, supplying countries would be able to make immediate application for funds for applied conservation and development projects that need to be undertaken urgently. Second, the money could be accessed directly, through existing United Nations environmental aid and donor programs, for example, without the need to generate dedicated systems of disbursement. Third, the portion of funds that is now spent on monitoring contractual agreements and tracing successive uses could also be reduced, releasing resources for other social purposes. I would not propose completely abandoning the project of monitoring the ways in which materials are used, as questions of how such materials are collected and used clearly have important social, cultural, and political implications. These tasks should continue, but they need not be seen as a necessary prerequisite for the release of compensation, as they are now.

Collecting institutions, companies, and research institutes would also benefit greatly from being released from the burden of having to account for every successive use that they make of collected material and from having to implement and monitor highly nuanced and complex access and benefit-sharing agreements. This would result in a substantial reduction of their administrative costs. The savings that are made might be used, in the corporate world, to perhaps cross-subsidize further research and development of basic drugs for use in developing countries. Scientific, academic, and other research institutes might well apply their savings to the pursuit of further basic research for which federal funds have been cut, such as species-identification programs.

All of these parties would also benefit immeasurably from a return to what might be thought of a "moderated free-access regime." The collection of species for taxonomic and other purposes would no longer be hindered, and students would be free to pursue crucial field research. This may draw attention to areas or species at risk of further destruction. Collectors from both developed and developing countries would be equally free to fully explore all the potential applications of biological materials and to work experimentally with those materials to produce new inventions and innovations. These may, in time, become marketable products that will provide a future revenue stream. Although such a scheme may seem too revisionist, it is more likely to fulfill the shared aim of finding new, creative ways to justly and equitably develop genetic and biochemical resources and information for the broader benefit of humankind than those that we have available to us today. While the difficulties of administering such a fund may seem insurmountable, it is difficult to imagine that they would exceed those associated with the current system. As radical as it seems, this new superfund model may well act to unblock some of the obstructions that now threaten to fatally impede the operation of existing regulatory and compensatory frameworks.

Undertaking this research project has involved immersing myself in a shifting debate about how best to govern the use of a new commodity: variously embodied types of bio-information. Although we continue to imagine that biological material is the same type of resource that it always has been, this detailed research, I hope, has revealed that it is in fact a resource that is being construed, represented, and utilized in new and unexpected ways. I have argued here that we must consequently learn to evolve sophisticated and custom-made systems of regulation that might reflect the complexities that attend its changing construction and use. My motivation in undertaking this research has not been to deride or undermine the inventive approaches to the regulation of this new commodity that have been developed in the last ten years but rather to build upon and learn from such endeavors. I am indebted to all those practitioners in the bioprospecting industry whose willingness to be frank about the shortcomings of the existing system has enabled us to reflect in a much more informed way on where and how our collective approach to the uses and regulation of this new resource might be improved.

By drawing out the parallels that exist between biotechnologies and other informational technologies, I hope that I broadened understandings of how biological materials are constructed and used as commodities in the contemporary life-sciences industry and revealed the capacity that this has for creating a new and highly lucrative resource economy in bio-information. I also hope that this analysis will lead us to reflect on questions of social and economic justice and of our individual obligations to ensure that this new resource economy operates in a just and equitable fashion. While the collection of plant and animal samples for use in the pharmaceutical industry may be declining, the collection and use of human tissue and organs is on the rise. The issues that I raise in this work—questions of what the status of various biological derivatives and "works" might be and of who should have access to them and under what terms and conditions—will have growing implications for us as the flourishing life-sciences industry pushes forward with the creation of DNA and stem-cell banks, DNA-sequence databases, and, indeed, enters other realms that remain, as yet, unimagined and uncharted. As we move into the twenty-first century, the century, perhaps, in which the engineering of life will become our central preoccupation, it seems appropriate that we continue to focus our collective attention on the question of how we, as a global community, wish to manage the commodification of life.

APPENDIX: METHODOLOGY

I have written elsewhere and in more detail about the process of undertaking this research,[1] so my aim here is simply to provide a brief summation of the research methodology. The task of establishing the size and composition of the bioprospecting industry required conventional and extensive methods of data collection—procuring statistics for collecting activities, establishing the number and type of companies involved, and so forth. The task of establishing how materials were being collected, concentrated, used, exchanged, or traded, how these processes were being controlled, and by whom, was clearly going to require a more inventive strategy. While a certain amount of information about the collecting activities of different companies could be acquired through searches of secondary sources (for example, published articles and company reports) it was apparent that these would not provide me with insights into how control over this trade was being exercised. In order to ascertain the kind of detailed and sensitive information that I required, I knew that I would have to conduct many interviews with those individuals who were ideally placed to comment on the nature and organization of this burgeoning trade.

I first selected a representative sample of companies and organizations that are centrally involved in bioprospecting for the pharmaceutical industry. These included a number of scientific and academic collecting institutes, large and small pharmaceutical companies, privately and publicly funded research agencies, biotechnology companies that specialize in the collection of genetic materials, individual brokers, conservation agencies, and NGOs. These organizations, companies, and institutions were not selected arbitrarily. They were chosen, after consultation with specialists in the field, because they have played a key role in the organization of this trade or because they are representative of those that have. The individuals that I chose to interview within these institutions were selected either because they have had a long or close association with bioprospecting programs; because they play a key role in the organization or day to day operation of this trade in genetic materials; or because they have devised compensatory agreements or can offer an informed perspective on their development. I conducted nearly forty interviews in the United States and

Britain for this project from 1995 through 2002. Most of the interviews were between 90 and 120 minutes in duration.

As the research progressed it became apparent that I required further information about the specific activities of scientific and academic collecting institutes. The sheer logistics of interviewing in a large number of institutions were such that I felt it would be more efficient to gather this information via a survey. My aims in undertaking the survey were to establish, first, whether there was an identifiable increase in attempts by commercial users to gain access to collections of biological materials held by scientific and academic institutions; second, what compensation was being offered to institutions in return—for example, offers of money or offers to underwrite the cost of future collecting programs; third, whether or not agreements had been entered into and compensation accepted; fourth, whether or not the collecting institution intended to share this compensation with supplying countries; and, finally, collecting institutions' motivation for allowing corporate access to their collections and their assessment of the impact that such activities might have on future relations with the countries in which they collect and on their own research agendas.

I approached the director of the association of Systematics Collections for permission to survey that organization's membership, as they constitute a representative sample of academic and scientific biological collection institutes in the United States. With that permission in hand, I contacted seventy-five member institutions of the ASC, their full complement, excepting those institutions that held paper records only. Forty-eight institutions responded, or 52 percent of the total, which was particularly high given the sensitive nature of the questions and the fact that the survey was administered by post. I am indebted to Elaine Hoagland and the membership of the ASC for allowing me to survey them in this way and for their willingness to discuss openly their concerns in relation to these matters.

Some of the most valuable primary data that I collected takes the form of corporate or institutional documents or reports that have been circulated between particular key actors but that have yet to enter the public domain or be comprehensively collated in a single site. Part of my research strategy was to collect as much of this documentary material as I could from the institutions or corporations in question during the course of my research. I had to embark on a lengthy process of negotiation in order to take photographs of collections or collection spaces. Interestingly, this proved even more contentious than asking for documents, and I was explicitly refused permission on almost every occasion. It was only after a protracted period of negotiation with Gordon Cragg of the NCI that I was allowed me to take photographs of the NCI repository. I am indebted to him for his generosity in assisting me with this research. I consider these photographs, which are entered here in the public domain for the first time, to make an important contribution to the public understanding of bio-prospecting practice.

NOTES

PREFACE

1. C. Price, "It Could Be Time to Face the Music: MP3 and Illicit Recordings," *The Financial Times*, 5 July 2000, 11.
2. E. Schierer, J. Bernoff, J. Sorley, M. and Gerson, *Content Out of Control: A Forrester TechStrategy Report* (Cambridge Mass.: Forester Publications, 2000), 8.
3. T. Radford, "The Breakthrough That Changes Everything," *The Guardian*, 26 June 2000.

1. INTRODUCTION

1. W. Reid et al., eds., *Biodiversity Prospecting: Using Genetic Resources for Sustainable Development* (Washington, D.C.: World Resources Institute, 1993), 1.
2. P. Dicken, *Global Shift: Transforming the World Economy*, 3d ed. (London: Chapman, 1999), 27–29
3. W. Bains, "Biotechnology for Pleasure and Profit" *Bio/Technology* 11 (November 1993): 1246.
4. MUD History, "Biotech Starts 2000 with a Bang As Recent Gains Rival Internet Stocks," (2000) available at http://www.bio.org/newsroom/newsitem.asp?id=2000_0120_01.
5. Biotechnology Industry Organization, "Report on Revenues of Selected Companies, 1980–2001," available at http://www.bio.org/investor/signs/200210rv5.asp.
6. MUD History, "Biotech Starts 2000 with a Bang."
7. Biotechnology Industry Organization, "Report on Biotech Revenues, 1992–2001," available at http://www.bio.org/investor/signs/200210rva.asp.
8. D. Haraway, *Modest_Witness@Second_Millennium.FemaleMan©_Meets_OncoMouse™: Feminism and Technoscience* (London: Routledge, 1997), 246.
9. K. Robins and D. Morley, *Spaces of Identity: Global Media, Electronic Landscapes, and Cultural Boundaries* (London: Routledge, 1995).
10. B. C. Parry, "The Fate of the Collections: Social Justice and the Annexation of Plant Genetic Resources," in *People, Plants and Justice: The Politics of Nature Conservation*, ed. C. Zerner (New York: Columbia University Press, 2000), 374–402, provides an earlier summation of this research.
11. This idea was first set out, in a tentative form at least, in J. Kloppenburg, *First the Seed: The Political Economy of Plant Biotechnology, 1492–2000* (Cambridge: Cambridge University Press, 1988) and alluded to in Manuel Castells, *The Informational City* (Oxford: Blackwell, 1989).

12. Castells, *The Informational City*, 12.

2. THE COLLECTION OF NATURE AND THE NATURE OF COLLECTING

1. P. Findlen, "Courting Nature," in *Cultures of Natural History*, ed. N. Jardine, J. Secord, and E. Sparry (Cambridge: Cambridge University Press, 1996), 57.

2. For a detailed description, see S. Bedini, *The Pope's Elephant* (Nashville: Sanders and Company, 1998); my thanks to Lisa Jardine for introducing me to "Hanno."

3. T. Eisner, "Prospecting for Nature's Chemical Riches," *Issues in Science and Technology* 6, no. 2. (Winter 1989): 31–34.

4. D. Lewis, "The Gene Hunters," *The Geographical Magazine* 63, no. 1 (January 1991): 36–43.

5. S. King, "Antifungal in the Jungle," *Worth*, Winter 1994/1995, 87.

6. King, "Antifungal in the Jungle," 87.

7. V. Shiva, *Monocultures of the Mind: Perspectives on Biodiversity and Biotechnology* (Penang, Malaysia: Third World Network Publishers, 1993); V. Shiva, "Why We Should Say 'No' to GATT TRIPS," *Third World Resurgence*, no. 39 (November 1993): 32–34; P. R. Mooney, "The Law of the Seed: Another Development in Plant Genetic Resources," *Development Dialogue*, nos. 1–2 (1983): 1–72; P. R. Mooney, "Exploiting Local Knowledge: International Policy Implications" in *Cultivating Knowledge: Genetic Diversity, Farmer Experimentation and Crop Research*, ed. W. De Boef et al. (London: Intermediate Technology Publications, 1993), 171–98. For a further discussion of "bio-piracy, "also see the communiqués of the Action Group on Erosion, Technology, and Concentration, formerly the Rural Advancement Foundation International, available at http://etcgroup.org/search.asp?type=communique.

8. See, for example P. Cox, "Ethnopharmacology and the Search for New Drugs," in *Ciba Foundation Symposium No. 154*, ed. D. J. Chadwick, and J. Marsh (Chichester: J. Wiley and Sons, 1990), 41; A. Sittenfield and A. Lovejoy, "Biodiversity Prospecting," *Our Planet* 6, no. 4 (1994): 20; N. Farnsworth, "Screening Plants for New Medicines" in *Biodiversity*, ed. E. O. Wilson (Washington, D.C.: National Academy Press, 1988), 83; E. Newman, "Earth's Vanishing Medicine Cabinet: Rain Forest Destruction and Its Impact on the Pharmaceutical Industry," *American Journal of Law and Medicine* 20, no. 4. (1994): 479.

9. J. Clifford, *The Predicament of Culture: Twentieth-Century Ethnography, Literature, and Art* (New York: Harvard Press, 1988), 218.

10. S. Pearce, *On Collecting: An Investigation into Collecting in the European Tradition* (London: Routledge, 1995), vii

11. Important exceptions to this are L. Brockway, *Science and Colonial Expansion: The Role of the British Royal Botanical Gardens* (New York: Academic Press, 1979); and J. R. Kloppenburg, *First the Seed: The Political Economy of Plant Biotechnology, 1492–2000* (Cambridge: Cambridge University Press, 1988).

12. See for example, D. Outram, "New Spaces in Natural History," in *Cultures of Natural History*, ed. N. Jardine, J. Secord, and E. Spary (Cambridge: Cambridge University Press, 1996), 249–65; S. Forgan, "Context, Image, and Function: A Preliminary Enquiry Into the Architecture of Scientific Societies," *The British Journal for the History of Science* 19 (1986): 89–113; and T. Markus, *Buildings and Power: Freedom and Control in the Origins of Modern Building Types* (New York: New Brunswick, 1993) .

13. *The Concise Oxford Dictionary of Current English*, 10th ed., s.v. "collection."

14. Bedini, *The Pope's Elephant*, 82.

15. L. Daston and K. Park, *Wonders and the Order of Nature, 1150–1750* (New York: Zone Books, 1988), 88.

16. Pearce, 189.

17. Carpenter, quoted in Clifford, *The Predicament of Culture*, 239; my italics.

18. Ibid., 223.

19. Ibid., 227. See also K. Pomian, *Collectors and Curiosities: Paris and Venice, 1500–1800* (Cambridge: Polity, 1990).

20. R. Belk, *Collecting in a Consumer Society* (London: Routledge, 1995).

21. Daston and Park, *Wonders and the Order of Nature*, 68.

22. S. Mullaney, "Strange things, Gross Terms, Curious Customs: The Rehearsal of Cultures in the Late Renaissance," *Representations* 3 (Summer 1983): 40.

23. Belk, *Collecting in a Consumer Society*, 32–34.

24. K. Whitaker, "The Culture of Curiosity," in *Cultures of Natural History*, ed. N. Jardine, J. Secord, and E. Spary (Cambridge: Cambridge University Press, 1996), 87.

25. K. Thomas, *Man and the Natural World: Changing Attitudes in England, 1500–1800* (London: Penguin, 1972), 27.

26. J. Gasgoine, "The Ordering of Nature and the Ordering of Empire: A Commentary," in *Visions of Empire: Voyages, Botany, and Representations of Empire*, ed. D. Miller and P. Reill (Cambridge: Cambridge University Press, 1996), 108.

27. For those interested in a more detailed and sophisticated reading of collecting practice in this period, see Daston and Park, *Wonders and the Order of Nature*; and P. Findlen, *Possessing Nature: Museums, Collecting, and Scientific Culture in Early Modern Italy* (Berkeley: University of California Press, 1994), both of which I have drawn on here extensively.

28. Findlen, *Possessing Nature*, 153.

29. Ibid., 158–59.

30. B. Latour, *Science in Action: How to Follow Scientists and Engineers Through Society* (Cambridge, Mass.: Harvard University Press, 1987), 215–57.

31. Ibid., 224.

32. Ibid.

33. Ibid., 219.

34. Ibid.

35. Ibid.

36. Ibid., 223.

37. J. Murdoch, "Inhuman/Nonhuman/Human: Actor-Network Theory and the Prospects for a Nondualistic and Symmetrical Perspective on Nature and Society," *Environment and Planning D* 15. no. 6 (1997): 741.

38. Herbarium sheets are specimens of plant material affixed to sheets of paper annotated with classificatory information.

39. It is, however, important to note that processes of classification came to rest upon comparisons of available material—those able to withstand transportation or capable of being affixed to a sheet of paper—a highly selective process in itself.

40. Herbaria are cases and/or rooms that contain archived herbarium sheets.

41. For a more detailed discussion of the development of the greenhouse and other glasshouse technology, see J. Hix, *The Glasshouse* (New York: Phaidon, 1996); and M. Woods and A. Warren, *Glasshouses: A History of Greenhouses, Orangeries, and Conservatories* (New York: Rizzoli, 1988).

42. See D. Massey and J. Allen, *Geography Matters!* (Cambridge: Cambridge University Press, 1984).

43. D. Miller, "Joseph Banks, Empire, and the 'Centres of Calculation' in Late Hanoverian London," in *Visions of Empire: Voyages, Botany, and Representations of Empire*, ed. D. Miller, and P. Reill (Cambridge: Cambridge University Press, 1996), 24.

44. Pearce, *On Collecting*, 255–56.

45. Clifford, *The Predicament of Culture*, 218.

46. Pearce, *On Collecting*, 267.

47. L. Schiebinger, "Gender and Natural History," in *Cultures of Natural History*, ed. N. Jardine, J. Secord, and E. Sparry (Cambridge: Cambridge University Press, 1996), 167.

48. Linnaeus himself later acknowledged that his sexual system of classification was a highly artificial one.

49. P. F. Stevens, *The Development of Biological Systematics: Antoine-Laurent de Jussieu, Nature, and the Natural System* (New York: Columbia University Press, 1994), 201.

50. Schiebinger, "Gender and Natural History," 171.

51. L. Koerner, "Carl Linnaeus in His Time and Place," in *Cultures of Natural History*, ed. N. Jardine, J. Secord, and E. Sparry (Cambridge: Cambridge University Press, 1996), 148.

52. D. Mackay, "Agents of Empire: The Banksian Collectors and Evaluation of New Lands," in *Visions of Empire: Voyages, Botany, and Representations of Empire*, ed. D. Miller and P. Reill (Cambridge: Cambridge University Press, 1996), 38.

53. E. B. Keeney, *The Botanists: Amateur Scientists in Nineteenth-Century America* (Chapel Hill: University of North Carolina Press, 1992), 10.

54. Stevens, *The Development of Biological Systematics*, 206.

55. Miller, "Joseph Banks," 24.

56. M. Foucault, *The Order of Things: An Archaeology of the Human Sciences* (New York: Vintage, 1970), 131.

57. Ibid.; my italics.

58. The sociologist Manuel Castells introduced the term a "space of flows" to alert people to the fact that even through they may remain largely invisible to us, the exchange and circulation of informational resources takes on a particular form and occupies a particular space within the global economy—a space of transmission or flow.

59. Quoted in J. Browne, "Biogeography and Empire," in *Cultures of Natural History*, ed. N. Jardine, J. Secord, and E. Sparry (Cambridge: Cambridge University Press, 1996), 313.

60. Ibid.

61. Pratt, M-L. *Imperial Eyes: Travel Writing and Transculturation* (London: Routledge, 1992), 31; my italics.

62. Castells's term refers here to the new spatial logic that characterizes the operation of an informational economy, one in which the flow of information, goods, and people between particular places takes on a greater significance than the places themselves.

63. Pearce, *On Collecting*, 302.

64. S. Fuller, "Studying the Proprietary Grounds of Knowledge," *Journal of Social Behaviour and Personality* 6, no. 6, special issue, "To Have Possessions: A Handbook on Ownership and Property," ed. F. W. Rudmin (1991): 105–28; quoted in Pearce, *On Collecting*, 302.

65. The pecuniary value of the entire collection or of the pieces within it is only determined when they are placed as saleable commodities within a circuit of exchange. This is as true of collections of cultural artifacts as it is of collections of natural materials. However, natural materials differ in that they are sometimes living and may therefore reproduce themselves, and/or mutate through breeding processes to create new materials that can, in turn, become the subject of a new collection processes.

66. Miller, "Joseph Banks," 8.

67. For a more detailed description of this relationship see Miller, "Joseph Banks," and Mackay, "Agents of Empire."

68. Pratt, *Imperial Eyes*, 25.

69. Miller, "Joseph Banks," 33. Also see A. E. Musson and E. Robertson, *Science and Technology in the Industrial Revolution* (Manchester: Manchester University Press, 1969).

70. R. Drayton, *Nature's Government: Science, Imperial Britain, and the "Improvement" of the World* (New Haven, Conn.: Yale University Press, 2000), 130

71. Ibid.

72. Quoted in ibid., 156–57.

73. Ibid.

74. Particularly the Dutch, the English, and the French.

75. Drayton, *Nature's Government*, 249.

76. Kloppenburg, *First the Seed*, 157.

77. Drayton, *Nature's Government*, 249.

78. See Brockway, "Science and Colonial Expansion," and Kloppenburg, "First The Seed," for a more detailed analysis.

79. Miller, "Joseph Banks," 54.

80. The name given to reproductive components of a plant, usually seeds or other materials from which plants can be propagated.

81. K. Lemmon, *Golden Age of Plant Hunters* (London: Phoenix House, 1968).

82. Kloppenburg, *First the Seed*, 161. In October 1994 the IARCs agreed to place their collections under the auspices of the Food and Agriculture Organisation (FAO). Under this agreement, developing-country germplasm remains available for public use; however, new restrictions prevent recipients from patenting this material. With the enactment of this policy, some 550,000 gene-bank accessions were placed firmly in the public domain. However, the agreement does not deal with materials derived from these accessions and, hence, remains problematic. More recently, scientists at the International Centre for Tropical Agriculture (CIAT) have taken the unprecedented step of threatening legal action against a U.S. company that had successfully claimed an exclusive patent on the enola bean—a traditional Mexican yellow bean. Although the bean in question was not taken from their accessions, the institution believes that the claim threatens public utilization of its identical yellow-bean accessions held under trust with the FAO.

3. SPEEDUP: ACCELERATING THE SOCIAL AND SPATIAL DYNAMICS OF COLLECTING

1. For a history of the practice of ethnobotany, see R. I. Ford, "Ethnobotany: Historical Diversity and Synthesis," in *The Nature and Status of Ethnobotany*, ed. R. I. Ford (Ann Arbor, Mich.: Museum of Anthropology Publication, 1978), 47–62; and for a more recent biography of the founder of the discipline see W. Davis, *One River: Science, Adventure, and Hallucinogens in the Amazon Basin* (London: Simon and Schuster, 1996). For a summation of work in this field, see N. Farnsworth, "Screening Plants for New Medicines," in *Biodiversity*, ed. E. O. Wilson, (Washington, D.C.: National Academy Press, 1988), 83–97; N. Farnsworth, "The Role of Ethnopharmacology in Drug Development," in *Proceedings of the Ciba Foundation Symposium: Bioactive Compounds From Plants*, ed. the Ciba Foundation (Chichester: Wiley, 1990), 2–21; M. Balick, "Ethnobotany and the Identification of Therapeutic agents From the Rainforest," in *Proceedings of the Ciba Foundation Symposium: Bioactive Compounds From Plants*, ed. the Ciba Foundation (Chichester: Wiley, 1990), 22–39; P. A. Cox,

"Ethnopharmacology and the Search For New Drugs," in *Proceedings of the Ciba Foundation Symposium: Bioactive Compounds From Plants*, ed. the Ciba Foundation (Chichester: Wiley, 1990), 40–55.

2. W. Reid et al., eds., *Biodiversity Prospecting: Using Genetic Resources for Sustainable Development* (Washington, D.C.: World Resources Institute, 1993), 1; my italics.

3. *The Oxford English Dictionary*, 2d ed., s.v. "resource."

4. C. Katz, "Whose Nature, Whose Culture? Private Productions of Space and the 'Preservation' of Nature," in *Remaking Reality: Nature at the Millennium*, ed. Braun and N. Castree (London: Routledge), 46–47.

5. P. R. Mooney, "The Law of the Seed: Another Development in Plant Genetic Resources," *Development Dialogue*, nos. 1–2 (1983): 1–72; and J. Doyle, *Altered Harvest: Agriculture, Genetics, and the Fate of the World's Food Supply* (New York: Viking, 1985).

6. See, for example, D. Bell, *The Coming of Post-Industrial Society: A Venture in Social Forecasting* (Harmondsworth: Penguin, 1974); and A. Touraine, *The Postindustrial Society* (London: Wildwood House, 1974).

7. M. Castells, *The Informational City* (Oxford: Blackwell, 1989), 13; my italics.

8. E. Fox Keller, *The Century of the Gene* (Cambridge, Mass.: Harvard University Press, 2000), 87

9. One extremely important exception is the work of the late historian of science Lilly Kay. See L. E. Kay, *Who Wrote the Book of Life: A History of the Genetic Code* (Stanford, Calif.: Stanford University Press, 2000).

10. D. Harvey, *Justice, Nature, and the Geography of Difference* (London: Basil Blackwell, 1996), 163.

11. Ibid.

12. D. Bohm, and D. Peat, *Science, Order, and Creativity* (London: Routledge, 1989), 39.

13. D. Haraway, *Symians, Cyborgs, and Women: The Re-invention of Nature* (New York: Routledge, 1991); D. Haraway, *Modest_Witness@Second_Millennium.FemaleMan©_Meets_OncoMouse™: Feminism and Technoscience* (New York: Routledge, 1997) ; L. Kay, "Who Wrote the Book of Life? Information and the Transformation of Molecular Biology, 1945–55," *Science in Context* 8, no. 4 (1995): 609–34; and E. Martin, "The End of the Body," *American Ethnologist* 19, no. 1 (1992): 121–40.

14. Kay, "Who Wrote the Book of Life?" 611.

15. Haraway, *Symians, Cyborgs, and Women*, 58.,

16. Kay, "Who Wrote the Book of Life?" 614.

17. Haraway, *Modest_Witness*, 245–46.

18. Kay, "Who Wrote the Book of Life?" 609.

19. Ibid., 610.

20. Ibid., 609.

21. Bell, *The Coming of Post-Industrial Society*; Touraine, *The Postindustrial Society*.

22. In works such as Bell, *The Coming of Post-Industrial Society*; Touraine, *The Postindustrial Society*; D. Lyon, *The Information Society: Issues and Illusions* (Oxford: Polity Press, 1988); P. Drucker, *Post-Capitalist Society* (Oxford: Butterworth-Heinnemen, 1993); H. Ernste and C. Jaeger, *Information Society and Spatial Structure* (London: Belhaven Press, 1989); W. Dutton, ed., *Information and Communication Technologies: Visions and Realities* (Oxford: Oxford University Press, 1996).

23. D. Harvey, *The Condition Of Postmodernity: An Enquiry into the Origins of Cultural Change* (London: Basil Blackwell, 1990), 159

24. David Harvey, telephone conversation with author, 6 September 1998. I am indebted to Professor Harvey for taking the time to discuss these matters with me at 11 A.M. on a public holiday. Any errors or misconceptions remain solely mine.

25. M. Strathern, "Potential Property: Intellectual Property Rights and Property in Persons," *Social Anthropology* 4, no. 1 (1996): 17–32.

26. B. Latour, *Science in Action: How to Follow Scientists and Engineers Through Society* (Cambridge, Mass.: Harvard University Press, 1987), 243.

27. Perhaps a better description here would be a de- and rematerialization—a rendering of the information in a new, more transmissible material form.

28. Castells, *The Informational City*, 12.

29. For an excellent, and more detailed, analysis of the Visible Human Project, see C. Waldby, *The Visible Human Project: Informatic Bodies and Posthuman Medicine* (London: Routledge, 2000).

30. Haraway, *Modest_Witness*, 244–55.

31. D. Hartl and W. Jones, *Essential Genetics* (Boston: Jones and Bartlett Publishers, 1999), 227.

32. "Where the subject of the obligation is a thing of a given class, the thing is said to be fungible, i.e. the delivery of any object which answers to the generic description will satisfy the terms of the obligation" (*Oxford English Dictionary*, 2d ed., s.v. "fungible").

33. W. Bains, "Biotechnology for Pleasure and Profit," *Bio/Technology* 11 (November 1993): 1246.

34. K. Leutwyler, "Send in the Clones," *Scientific American*, 27 July 1998, available at http://www.sciam.com/article.cfm?articleID=000186A6-697C-1CE2-95FB809EC588EF21.

35. Waldby, *The Visible Human Project*, 7.

36. J. H. Barton, "The Biodiversity Convention and the Flow of Scientific Information" in *Global Genetic Resources: Access, Ownership, and Intellectual Property Rights*, ed. K. E. Hoagland and A. Y. Rossman (Washington, D.C.: The Beltsville Symposia in Agricultural Research/The Association of Systematics Research Publishers, 1997), 55.

37. J. R. Kloppenburg, *First the Seed: The Political Economy of Plant Biotechnology, 1492–2000* (Cambridge: Cambridge University Press, 19880.

38. D. Goodman and M. Redclift, *Refashioning Nature: Food, Ecology, and Nature* (London: Routledge, 1991); and D. Goodman and J. Wilkinson, "Patterns of Research and Innovation in the Modern Agri-Food Sector," in *Technological Change and the Rural Environment*, ed. P. Lowe, T. Marsden, and S. Whatmore (London: David Fulton Publishers, 1990), 127–46.

39. Goodman and Redclift, *Refashioning Nature*, 191.

40. P. R. Mooney, *Seeds of the Earth: A Private or Public Resource?* (Ottawa: Pares Publishing, 1979); and G. Wilkes, "The World's Crop Plant Germplasm: An Endangered Resource," *Bulletin of Atomic Scientists* 33 (1977): 8–16.

41. P. Chaterjee and M. Finger, *The Earth Brokers: Power, Politics, and World Development* (London: Routledge, 1994), 144.

42. H. Daly and R. Goodland, "An Ecological Assessment of Deregulation of International Commerce Under GATT," Environmental Working Paper Draft Document prepared for The World Bank Environment Department (Geneva: The World Bank, 1993), 5.

43. The purpose of public disclosure is to foster technological innovation by enabling those "skilled in the art" to reproduce and improve upon inventions that might otherwise have remained trade secrets. The length of the term of patent protection was recently increased from seventeen to twenty years.

44. J. B. Thompson, *Studies in the Theory of Ideology* (Cambridge: Polity Press, 1984), 33.

45. J. Locke, *Two Treatises of Government*, ed. P. Laslett (1690; reprint, Cambridge: Cambridge University Press, 1988), 27.

46. For a more detailed analysis see P. Drahos, *A Philosophy of Intellectual Property* (Aldershot: Dartmouth, 1996).

47. W. R. C. Cornish, *Intellectual Property: Patents, Copyright, Trade Marks, and Allied Rights* (London: Sweet and Maxwell, 1999), 111.

48. K. Boehm, *The British Patent System*, vol. 1, *Administration* (Cambridge: Cambridge University Press, 1967), 22–30.

49. L. Bentley and B. Sherman, *Intellectual Property Law* (Oxford: Oxford University Press, 2001), 2.

50. Bentley and Sherman, *Intellectual Property Law*, 79.

51. The purpose of the disclosure by description clause is to enable those who follow to replicate or repeat the invention. This presented difficulties for breeders, as, unlike inventors of inanimate objects, they could not be certain that their invention would be reproduced in exactly the same way each time and would not evolve in time in such as way as to render the description inaccurate.

52. Established in *Funk Brothers Seed Co. v Kalo Inoculant Co.*, 333 U.S. 127 (1948).

53. *Diamond v Chakrabarty*, 447 U.S. 303.

54. Cited in *Diamond v Chakrabarty*, 309.

55. Ibid., 320.

56. Ibid.

57. Cited in V. Shiva, "Why We Should Say 'No' to GATT –TRIPs," *Third World Resurgence* no. 39 (November 1993): 33.

58. C. Juma, *The Gene Hunters* (London: Zed Books, 1989), 165.

59. See K. Aoki, "Intellectual Property and Sovereignty: Notes Towards a Cultural Geography of Authorship," *Stanford Law Review* 48 (1996): 1293–1355; J. Boyle, *Shamans, Software, and Spleens: Law and the Construction of the Information Society* (Cambridge, Mass.: Harvard University Press, 1996); R. Coombe, "Authorial Cartographies: Mapping Proprietary Borders in a Less-Than-Brave New World," *Stanford Law Review* 48 (1996): 1357–66.

60. W. Gordon, "A Property Right in Self-Expression: Equality and Individualism in the Natural Law of Intellectual Property," *Yale Law Journal* 102 (1993): 1533.

61. Bentley and Sherman, *Intellectual Property Law*, 415–17.

62. Shiva, "Why We Should Say 'No' to GATT–TRIPs," 33.

63. Bentley and Sherman, *Intellectual Property Law*, 381.

64. Sterrckx, S. "Some Ethically Problematic Aspects of the Proposal for a Directive on the Legal Protection of Biotechnological Inventions," *European Intellectual Property Review* 123 (1998): 125.

65. Kloppenburg, *First the Seed*, 263–64.

66. Ibid.

67. The history of the role of the Patent Office technicians in creating new means of regulating the use of biological materials would make a fascinating study but unfortunately remains beyond the scope of this work. It is worth noting, however, that that the demand for patent examiners for biotechnological inventions increased so rapidly in the late 1980s that the U.S. Patent Office was forced to increase their number from 91 to 112 in October 1988 and from 140 to 200 two years later. This was a necessary response to the increase in number and type of biotechnological inventions which saw the class C12N 15/00, "mutation or genetic engineering," go from a single undivided category of examination in 1979, to one divided into 90 subclasses by 1989. See P. Power, "Interaction Between Biotechnology and the Patent System," *The Australian Intellectual Property Law Journal* 3, no. 4 (1992): 218.

68. For example, within a decade of the *Diamond v. Chakrabarty* decision, the European Patent Office guidelines determined that while "to find a substance freely occurring in nature is mere discovery and therefore not patentable — if a substance found in nature has first to be *isolated* from it's surroundings [rescued from the chaos] . . . if it is 'new' in the absolute sense of having no 'previously recognized existence' [within Western systems of knowledge classification] then the substance *per se* may be patentable." The Japanese patent office later extended this

definition further to allow claims on chemical compounds contained in natural substances *in any instance* where "taking possession" and confirmation of the structure of the product could not be effected without artificial aids. For a more detailed discussion see S. Bent et al., *IPR in Biotechnology World-wide* (New York: Stockton Press, 1987).

69. M. Featherstone, "Global Culture: An Introduction," in *Global Culture: Nationalism, Globalisation, and Modernity*, ed. M. Featherstone (London: Sage Publications, 1990), 2.

70. Which in this instance can be seen as one and the same.

71. V. Shiva, *Monocultures of the Mind: Perspectives on Biodiversity and Biotechnology* (Penang, Malaysia: Third World Network Publishers, 1993).

72. P. R. Mooney, "Exploiting Local Knowledge: International Policy Implications," in *Cultivating Knowledge: Genetic Diversity, Farmer Experimentation, and Crop Research*, ed. W. De Boef et al. (London: Intermediate Technology Publications, 1993), 171–98.

73. Shiva, *Monocultures of the Mind*, 9.

74. Ibid., 82. Shiva does not strictly define these terms. Edward Said has suggested that "imperialism" is "a word and an idea today so controversial, so fraught with all sorts of questions, doubts, polemics and ideological premises as nearly to resist use altogether." Given this, I can only follow his lead and quote M. W. Doyle, *Empires* (Ithaca, N.Y.: Cornell University Press, 1986), 45, who describes the terms thus: "empire is a relationship, formal or informal, in which one state controls the effective political sovereignty of another political society. It can be achieved by force, by political collaboration, by economic, social or cultural dependence. Imperialism is the process or policy of establishing or maintaining empire" Said goes on to suggest that "in our time direct colonialism has largely ended, imperialism, as we shall see, lingers where it has always been, in a kind of general cultural sphere, as well as in specific political, ideological, economic and social practices." See E. Said, *Culture and Imperialism* (London: Vintage, 1993), 8.

75. See for example L. Glowka, "A Guide to the Convention on Biological Diversity" (New York: UNEP, 1993).

76. Ibid., 2.

77. World Commission on Environment and Development, *Our Common Future* (Oxford: Oxford University Press, 1987), 147.

78. N. Middleton, P. O'Keefe, and S. Moyo, *Tears of the Crocodile: From Rio to Reality in the Developing World* (London: Pluto Press, 1993), 55.

79. N. Hildyard, "Foxes in Charge of the Chickens," in *Global Ecology: A New Area of Political Conflict*, ed. W. Sachs (London: Zed Books, 1995), 22.

80. Chaterjee and Finger, *The Earth Brokers*, 42.

81. Ibid., 70.

82. Determined under *The International Undertaking on Plant Genetic*

Resources Res. 8/83, F.A.O. Conference 22d Sess., Annex to Res. 8/83 at Art. 2., F.A.O. Doc. C83/REP (1983).

83. The Convention on Biological Diversity, 5 June 1992, Article 2, 109. For a more detailed discussion of the potentially conflictual relationship between the Biodiversity Convention and the GATT TRIPS agreement, see D. Downes, "The Convention on Biological Diversity and the GATT," in *The Use of Trade Measures in Select Multilateral Environmental Agreements*, ed. R. Housman et al. (Nairobi, Kenya: UNEP Publishers, 1995), 197–251.

84. D. Pember, *Mass Media Law* (Chicago: Brown and Benchmark Publishers, 1997), 449.

85. W. Overbeck, *Major Principles of Media Law* (Philadelphia: Harcourt Brace College Publishers, 1996), 210.

86. Ibid., 211.

87. E. Simon, "Innovation and Intellectual Property Protection: The Software Industry Perspective," *The Columbia Journal of World Business* 31, no. 1 (1996): 33.

88. A. Johnson-Laird, "Exploring the Information Superhighway: The Good, The Bad, and the Ugly," paper presented at the Electronic Information Law Institute Meeting, San Francisco, 2–3 March 1995.

89. *Georgia Television Company v Television News Clips of Atlanta*, 983 F. 2d. 238, (2d Cir. 1993).

90. Pember, *Mass Media Law*, 480.

91. Ibid.

92. The question of whether forms of copyright might usefully be employed to control access to and use of biological "works" remains the subject of ongoing debate. See, for example, G. Moore and A. Wilson, "Pulling on the Genes," *IT Law Today* 10, no. 3 (2002): 1–13; S. Connor, "DNA Codes May Be Protected as 'Music,' " *The Independent* (London) 22 March 2002.

4. NEW COLLECTORS, NEW COLLECTIONS

1. K. ten Kate and S. Laird, *The Commercial Use of Biodiversity: Access to Genetic Resources and Benefit-Sharing* (London: Earthscan, 1999), 1.

2. Dianne De Furia, director of commercial development at Bristol Myers Squibb, interview with author, tape recording, Princeton, N.J., 24 June 1996.

3. Ibid.

4. N. Farnsworth, "Screening Plants for New Medicines," in *Biodiversity*, ed. E. O. Wilson (Washington, D.C.: National Academy Press, 1988), 83–97.

5. J. Berdy, "The Discovery of New Antibiotic Microbial Metabolites: Screening and Identification," *Progress In Microbiology* 27 (1989): 3–25.

6. J. McChesney, "Biological Diversity, Chemical Diversity and the Search for New Pharmaceuticals," paper presented at the Symposium on Tropical Forest Medical Resources and the Conservation of Biodiversity, Rainforest Alliance, New York, 1992.

7. J. Di Masi, et al., "Cost of Innovation in the Pharmaceutical Sector," *Journal of Health Economics* 10 (1991): 107–42.

8. Robert Borris, senior research fellow, Merck and Co., interview with author, tape recording, Rahway, N.J., 29 April 1996.

9. Ibid.

10. Ibid.

11. Henry Shands, national program leader of the USDA Agricultural Research Service, interview with author, tape recording, Beltsville, Md., 21 March 1996.

12. The NIH comprises five separate research institutes. The National Cancer Institute and the National Institute of Allergies and Infectious Diseases are the two that are most interested in natural-products research.

13. Gordon Cragg, director of the Natural Products Branch of the National Cancer Institute, interview with author, tape recording, Bethesda, Md., 8 August 1995.

14. Such as chemical or pharmaceutical companies or academic research institutions.

15. A smaller collection and screening program had been carried out in 1955 in collaboration with the pharmaceutical industry. This was limited in scale, targeting only microorganisms, bacteria, and fungi.

16. Cragg, interview.

17. Rural Advancement Foundation International, "Bioprospecting and Indigenous Peoples: An Overview," paper prepared for the South Pacific Consultation on Indigenous People's Knowledge and Intellectual Property Rights, United Nations Development Program, Fiji, 24–27 April 1995, 2.

18. M. Boyd and K. Paull, "Some Practical Considerations and Applications of the National Cancer Institute In Vitro Anticancer Drug Discovery Screen," *Drug Development Research* 34 (1995): 92; and G. Cragg, personal communication with author, 8 August 1995.

19. Cragg, interview.

20. Ibid.

21. Only thirty-nine countries are listed in the table, although this is necessarily a conservative estimate. The Coral Reef Foundation, for example, has undertaken collections on behalf of the NCI not only in Palau, but also in Fiji, PNG, the Philippines, Hong Kong, the Northern Marianas, Bahrain, and Zanzibar. The same is possibly true of other collecting organizations.

22. For a detailed analysis of this agreement, see W. Reid et al., eds., *Biodiversity Prospecting: Using Genetic Resources for Sustainable Development* (Washington, D.C.: World Resources Institute, 1993).

23. Shaman Pharmaceuticals, "Annual Report of 1994," 27.

24. Steven King, vice president of Shaman Pharmaceuticals, interview with author, tape recording, San Francisco, 28 September 1995.

25. Neil Belson, chief executive officer, Pharmacognetics, telephone interview with author, tape recording, 10 May 1996.

26. Malcom Morville, chief executive officer of Phytera, telephone interview with author, tape recording, 8 May 1996.

27. Promotional material supplied from the Knowledge Recovery Foundation, New York.

28. Rural Advancement Foundation International, "Bioprospecting and Indigenous Peoples," 2.

29. See S. Crooke, "Optimizing the Impact of Genomics on Drug Discovery and Development" *Nature Biotechnology* 16 (1998): S2.

30. *The Economist*, "Horn of Plenty: New Genetic Knowledge Means More and More Effective Drugs," 21 February 1998, S5.

31. Ten Kate and Laird, *The Commercial Use of Biodiversity*, 52.

32. C. Angerhofer and J. Pezzuto, "Application of Biotechnology in Drug Discovery and Evaluation," in *Biotechnology and Pharmacy*, ed. J. Pezzuto, M. Johnson, and H. Manase (London: Chapman and Hall, 1993), 312–65.

33. Boyd and Paull, "Some Practical Considerations," 91–109.

34. Cragg, interview.

35. Ibid.

36. G. Cragg and P. Boyd, "Drug Discovery and Development at the National Cancer Institute: The Role of Natural Products of Plant Origin," in *Medicinal Resources of the Tropical Forest: Biodiversity and Its Importance to Human Health*, ed. M. Balick, E. Elisabetsky, and S. Laird (New York: Columbia University Press, 1996), 101.

37. Doug Daly, interview with author, tape recording, New York, 13 March 1996.

38. M. Grever et al., "The National Cancer Institute: Cancer Drug Discovery and Development Program," *Seminars in Oncology* 19, no. 6 (1992): 622–38.

39. See, for example: W. K. Stevens, "Costa Rica in Pact to Search for Forest Drugs," *New York Times*, 24 September 1991; W. Booth, "U.S. Drug Firm Signs Up to Farm Tropical Forests," *Washington Post*, 21 September 1991; E. Blum, "Making Biodiversity Conservation Profitable: A Case Study of the Merck–InBio Agreement," *Environment* 35, no. 4. (May 1993): 16–29; L. Roberts, "Chemical Prospecting: Hope for Vanishing Ecosystems?" *Science* 256, no. 5060 (May 1992): 1142–43.

40. These agreements have mutated somewhat over time. They have become more complex, and their wording has changed accordingly. However, they remain the same in principle. The latest version of the MOU can be found at http://ttb.nci.nih.gov/npmou.html. The most recent "Letter of Collection" agreement is also available at http://www-otd.nci.nih.gov/nploc.html.

41. Grantees include all those agencies and institutions that are funded by the NCI's Natural Products Research Program, including those that collect on behalf of the NCI.

42. Cragg, interview.

43. The exception was the promotional material for Pfizer's collecting program,

which was launched in 1993. It made no mention of such issues, as they claimed that they were only intending to collect material within the mainland United States.

44. S. King, "Antifungal in the Jungle," *Worth*, Winter 1994/1995, 87.

45. A. Sittenfield and R. Villers, "Costa Rica's InBIO: Collaborative Biodiversity Research Agreements with the Pharmaceutical Industry," in *Principles of Conservation Management*, ed. G. K. Meffe and C. R. Carroll (Sunderland, Mass.: Sinaver, 1993), 500.

46. R. Gamez, "A New Lease on Life," in *Biodiversity Prospecting: Using Genetic Resources for Sustainable Development*, ed. W. Reid et al. (Washington, D.C.: World Resources Institute, 1993), 30.

47. Borris, interview.

48. See for example, Elisabetsky, quoted in C. Joyce, "Western Medicine Men Return to the Field," *Bioscience* 42, no. 6 (1992): 399–403; D. Gershon, "If Biodiversity Has a Price, Who Sets It and Who Should Benefit?" *Nature* 279 (1992): 284–95; A. Cunningham, *Ethics, Ethnobotanical Research, and Biodiversity* (Geneva: WWF International, Gland Publicatons, 1993); D. A. Posey, "Intellectual Property Rights and Just Compensation for Indigenous People," *Anthropology Today* 6, no. 4 (1990): 13–16.

49. B. Timmermann and B. Hutchinson, "Biodiversity Prospecting in the Drylands of Chile, Argentina, and Mexico," *Arid Lands Newsletter* no. 37 (Spring/Summer 1995): 12.

50. PhrMA represents the United States' leading research-based pharmaceutical and biotechnology companies, including all of the country's largest pharmaceutical companies.

51. G. Mossinghoff and T. Bombelles, "The Importance of Intellectual Property Protection to the American Research-Intensive Pharmaceutical Industry," *The Columbia Journal of World Business* 31, no. 1 (Spring 1996): 43

52. Ibid., 39.

53. Michael Balick, interview with author, tape recording, New York, 1 August 1995.

54. Kate Duffy Mazan, interview with author, tape recording, Bethesda, Md., 9 August 1995.

55. For a review of the establishment of the ICBG program, see F. T. Grifo, "Chemical Prospecting: An Overview of the International Biodiversity Co-operative Groups Program," in *Emerging Connections: Biodiversity, Biotechnology, and Sustainable Development in Health and Agriculture*, ed. J. Feinsilver (Washington, D.C.: Pan-American Health Organization, 1994), 12–16; J. Rosenthal, "Equitable Sharing of Biodiversity Benefits: Agreements on Genetic Resources," paper prepared for the OECD International Conference on Biodiversity Incentive Measures, 25–28 March 1996, Cairns, Australia; and Timmermann and Huchinson, "Biodiversity Prospecting in the Drylands of Chile, Argentina and Mexico." For a more nuanced and detailed ethnography of one of the ICBG programs, see C. Hayden, *When Nature Goes Pub-*

which was launched in 1993. It made no mention of such issues, as they claimed that they were only intending to collect material within the mainland United States.

44. S. King, "Antifungal in the Jungle," *Worth*, Winter 1994/1995, 87.

45. A. Sittenfield and R. Villers, "Costa Rica's InBIO: Collaborative Biodiversity Research Agreements with the Pharmaceutical Industry," in *Principles of Conservation Management*, ed. G. K. Meffe and C. R. Carroll (Sunderland, Mass.: Sinaver, 1993), 500.

46. R. Gamez, "A New Lease on Life," in *Biodiversity Prospecting: Using Genetic Resources for Sustainable Development*, ed. W. Reid et al. (Washington, D.C.: World Resources Institute, 1993), 30.

47. Borris, interview.

48. See for example, Elisabetsky, quoted in C. Joyce, "Western Medicine Men Return to the Field," *Bioscience* 42, no. 6 (1992): 399–403; D. Gershon, "If Biodiversity Has a Price, Who Sets It and Who Should Benefit?" *Nature* 279 (1992): 284–95; A. Cunningham, *Ethics, Ethnobotanical Research, and Biodiversity* (Geneva: WWF International, Gland Publicatons, 1993); D. A. Posey, "Intellectual Property Rights and Just Compensation for Indigenous People," *Anthropology Today* 6, no. 4 (1990): 13–16.

49. B. Timmermann and B. Hutchinson, "Biodiversity Prospecting in the Drylands of Chile, Argentina, and Mexico," *Arid Lands Newsletter* no. 37 (Spring/Summer 1995): 12.

50. PhrMA represents the United States' leading research-based pharmaceutical and biotechnology companies, including all of the country's largest pharmaceutical companies.

51. G. Mossinghoff and T. Bombelles, "The Importance of Intellectual Property Protection to the American Research-Intensive Pharmaceutical Industry," *The Columbia Journal of World Business* 31, no. 1 (Spring 1996): 43

52. Ibid., 39.

53. Michael Balick, interview with author, tape recording, New York, 1 August 1995.

54. Kate Duffy Mazan, interview with author, tape recording, Bethesda, Md., 9 August 1995.

55. For a review of the establishment of the ICBG program, see F. T. Grifo, "Chemical Prospecting: An Overview of the International Biodiversity Co-operative Groups Program," in *Emerging Connections: Biodiversity, Biotechnology, and Sustainable Development in Health and Agriculture*, ed. J. Feinsilver (Washington, D.C.: Pan-American Health Organization, 1994), 12–16; J. Rosenthal, "Equitable Sharing of Biodiversity Benefits: Agreements on Genetic Resources," paper prepared for the OECD International Conference on Biodiversity Incentive Measures, 25–28 March 1996, Cairns, Australia; and Timmermann and Huchinson, "Biodiversity Prospecting in the Drylands of Chile, Argentina and Mexico." For a more nuanced and detailed ethnography of one of the ICBG programs, see C. Hayden, *When Nature Goes Pub-*

lic: *The Making and Unmaking of Bio-Prospecting in Mexico* (Princeton, N.J.: Princeton University Press, 2003).

56. From a promotional release accompanying the launch of the ICBG program (Washington, D.C.: N.I.H./U.S. Aid/ National Science Foundation Publishers, 1993).

57. Ibid.

58. The ICBG program received $2.7 million over a five-year period. The successful consortia were chosen after a process of peer review. Reviewers were reported to come "from universities, museums, pharmaceutical companies, the World Bank, environmental non-profits with backgrounds in natural products chemistry, intellectual property rights law, systematics, ecology, ethnobiology and international development." For a fuller description of this process see Griffo, "Chemical Prospecting."

59. National Institutes of Health, National Institute of Mental Health, National Science Foundation, and the U.S. Agency for International Development, "International Cooperative Biodiversity Groups Agreement," 12 June 1992, Washington D.C.

60. Ibid.

61. Grifo, "Chemical Prospecting," 9.

62. P. Reilly, "Press Release from Conservation International: Forest People Search for New Medicines; Initiatives Builds Conservation-Based Industry," 7 December 1993, Washington, D.C., 4.

63. The National Museum of Natural History, Smithsonian Institution, *Annual Report* (Washington, D.C., 1995), 7.

64. Personal communication with Dr. Roger Waigh, Department of Pharmaceutical Sciences, University of Strathclyde, 4 January 1996. For a more recent update on the entrepreneurial activities of Strathclyde University, see D. MacLeod, "Survival of the Fittest," *The Guardian*, 19 May 1998.

65. Cragg, interview. Also see G. Cragg et al., "Natural Product Drug Discovery and Development at the National Cancer Institute: Policies for International Collaboration and Compensation," in *Monographs in Systematic Botany from the Missouri Botanical Garden*, vol. 48 (St. Louis: Missouri Botanical Gardens Press, 1994), 223

66. C. Joyce, "Prospectors for Tropical Medicines," *New Scientist*, 19 October 1991, 36–40.

67. Ibid.

68. R. Gamez, quoted in L. Tangley, "Cataloguing Costa Rica's Diversity," *Bioscience* 40, no. 9 (1990): 636.

69. See Joyce, "Prospectors for Tropical Medicines," 38

70. P. A. Cox, "Ethnopharmacology and the Search For New Drugs," in *Proceedings of the Ciba Foundation Symposium: Bioactive Compounds From Plants*, ed. the Ciba Foundation (Chichester: Wiley, 1990), 42.

71. N. Farnsworth, "The Role of Ethnopharmacology in Drug Development," in *Proceedings of the Ciba Foundation Symposium: Bioactive Compounds From Plants*, ed. the Ciba Foundation (Chichester: Wiley, 1990), 5; and M. Balick, "Ethnobotany and the Identification of Therapeutic agents From the Rainforest" in *Proceedings of the Ciba Foundation Symposium: Bioactive Compounds From Plants*, ed. the Ciba Foundation (Chichester: Wiley, 1990), 27.

72. Harshberger, quoted in C. M. Cotton, *Ethnobotany: Principles and Applications* (Chichester: Wiley, 1996), 1.

73. See Balick, "Ethnobotany and the Identification of Therapeutic Agents."

74. Cox, "Ethnopharmacaology and the Search for New Drugs," 44.

75. S. Greene, "Intellectual Property, Resources, or Territory? Reframing the Debate over Indigenous Rights, Traditional Knowledge, and Pharmaceutical Bioprospection," in *Truth Claims: Representation and Human Rights*, ed. Mark Philip Bradley and Patrice Petro. (New Brunswick, N.J.: Rutgers University Press, 2002), 229–49; C. Hayden, "From Market to Market: Bioprospecting's Idioms of Inclusion," *American Ethnologist* 30, no. 3 (2003): 359–71.

76. I. Anderson, "Oceans Plundered in the Name of Medicine," *New Scientist* 148, no. 2005 (1995): 5.

77. Cragg, G., interview with author, tape recording, Bethesda, Md., 19 March 1996.

78. Anderson, "Oceans Plundered, " 5.

79. T. McCloud et al., "Extraction of Bioactive Molecules from Plants," paper presented at The International Congress on Natural Products Research, Park City, Utah, 17–21 July 1988, 1, my italics; and T. McCloud et al., "Extraction of Bioactive Molecules from Marine Organisms," paper presented at The International Congress on Natural Products Research, Park City, Utah, 17–21 July 1988.

80. R. Hanner, curatorial associate, Ambrose Monell Cryo Collection at the American Museum of Natural History, telephone interview with author, 12 July 2002.

81. Ibid.

82. E. Rowe, Wessington Cryogenics, telephone interview with author, 10 August 2001.

83. B. Neirmann, director, Program in Molecular Biology and Virology, The American Type Culture Collection, interview with author, tape recording, Rockville, Md., 20 March 1996.

84. Balick, interview.

85. F. T. Griffo, interview with author, tape recording, New York, 28 March 1996.

86. Borris, interview.

87. Ibid.

5. THE FATE OF THE COLLECTIONS

1. G. Cragg, R. Boyd, M. Grever, and S. Schepartz, "Policies for International

Collaboration and Compensation in Drug Discovery and Development at the United States National Cancer Institute: The NCI Letter of Collection," in *Intellectual Property Rights for Indigenous Peoples: A Sourcebook*, ed. T. Greaves (Oklahoma City: The Society for Applied Anthropology, 1994), 89, my italics.

2. G. Cragg, interview with author, tape recording, Bethesda, Md., 8 August 1995.

3. See N. Farnsworth, "The Role of Ethnopharmacology in Drug Development," in *Proceedings of the Ciba Foundation Symposium: Bioactive Compounds from Plants*, ed. the Ciba Foundation (Chichester: Wiley, 1990), 2–21.

4. The "low-volume" in this case refers to the amount of chemical compounds required from the biomass to produce the drug rather than volume of sales of the drugs themselves. See J. Kloppenburg, *First the Seed: The Political Economy of Plant Biotechnology, 1492–2000* (Cambridge: Cambridge University Press, 1988), 276.

5. S. Brush, "A Non-Market Approach to Protecting Biological Resources," in *Intellectual Property Rights for Indigenous Peoples: A Sourcebook*, ed. T. Greaves (Oklahoma City: The Society for Applied Anthropology, 1994), 137.

6. Ibid.

7. F. Grifo, interview with author, tape recording, New York, 28 March 1996.

8. S. Laird, "Contracts for Biodiversity Prospecting," in *Biodiversity Prospecting: Using Genetic Resources for Sustainable Development*, ed. W. Reid et al. (Washington, D.C.: World Resources Institute, 1993), 118.

9. M. Balick, interview with author, tape recording, New York, 1 August 1995.

10. N. Farnsworth, "Screening Plants for New Medicines," in *Biodiversity*, ed. E. O. Wilson (Washington, D.C.: National Academy Press, 1988), 95.

11. McGraw-Hill Higher Education, "More Accessible Plant Alkaloids Through Cell Culture Systems," February 2000, available at http://www.mhhe.com/biosci/pae/botany/botany_map/articles/article_05.html.

12. R. Borris, interview with author, tape recording, Rahway, N.J., 29 April 1996; my italics.

13. Ibid.; my italics.

14. Ibid.; also, S. King, interview with author, tape recording, San Francisco, 28 September 1995.

15. M. C. Wani et al., "Plant Antitumor Agents. Part 6: The Isolation and Structure of Taxol, a Novel Antileukemic and Antitumor Agent from *Taxus brevifolia*," *Journal of the American Chemistry Society* 93, no. 9 (1971): 2325.

16. For a more comprehensive history, see J. Goodman and V. Walsh, *The Story of Taxol: Nature and Politics in the Pursuit of an Anti-Cancer Drug* (Cambridge: Cambridge University Press, 2001).

17. G. Cragg and P. Boyd, "Drug Discovery and Development at the National Cancer Institute: The Role of Natural Products of Plant Origin," in *Medicinal Resources of the Tropical Forest: Biodiversity and Its Importance to Human Health*, ed.

M. Balick, E. Elisabetsky, and S. Laird (New York: Columbia University Press, 1996), 117.

18. King, interview.

19. For a commentary on BMS's successful attempt to trademark the name Taxol, see *Nature*, "Names for Hi-jacking," editorial, *Nature* 373, no. 6513 (February 1995): 370.

20. For a fuller report, see C. Venkat, "Paclitaxel Production Through Plant Cell Culture: An Exciting Approach to Harnessing Biodiversity," paper presented at the International Conference on Biodiversity and Bioresources: Conservation and Utilization, Phuket, Thailand, 23–27 November 1997, available at http://www.iupac.org/symposia/proceedings/phuket97/venkat.html.

21. Phyton, Inc. "Phyton Expands Commercial Partnership with Bristol Meyers Squibb for Paclitaxel Supply," 1 July 2002, available at: http://www.phyton-inc.com/frame_news7-10-02.html.

22. Ibid.

23. G. Cragg, interview with author, tape recording, Bethesda, Md., 19 March 1996.

24. *The Economist*, "Biotech's Secret Garden" *The Economist*, 30 May 1998, 75.

25. M. Castells, *The Informational City* (Oxford: Blackwell, 1989), 13; my italics.

26. Description courtesy of the University of Louisville, Peptide Research Laboratory Web site: http://www.louisville.edu/a-s/chemistry/peptide/combichem.html.

27. R. Sykes, "Research: A Key Factor for the Development of Countries," 1 December 1999, available at http://www.fcs.es/fcs/eng/eidon/Introing/Eidon3/documentos/documentos.html.

28. *The Economist*, "From Blunderbuss to Magic Bullet," *The Economist*, 21 February 1998, S9.

29. Glaxo Wellcome, "More and Better Candidate Medicines from Glaxo Wellcome R&D," Glaxo Press Release Archive, London, 9 December 1997, available at http://www.gsk.com/press_archive/gw/1997/press_971209pr.htm.

30. R. Thomas, interview with author, tape recording, Sussex, UK, 1 Febraury 1996.

31. For a fuller discussion see B. Latour, *We Have Never Been Modern* (London: Harvester Wheatsheaf, 1993); and M. Serres, *Statues* (Paris: François Bourin, 1987).

32. Cragg, interview, 19 March 1996.

33. N. Belson, telephone interview with author, tape recording, 10 May 1996.

34. King, interview.

35. Cragg, interview, 19 March 1996.

36. D. De Furia, interview with author, tape recording, Princeton, N.J., 24 June 1996.

37. Borris, interview.

38. De Furia, interview; my italics.

39. Borris, interview.

40. De Furia, interview.

41. King, interview.

42. Borris, interview.

43. Belson, interview.

44. Cragg, interview, 19 March 1996.

45. J. Kress, interview with author, tape recording, Washington, D.C., 21 April 1996.

46. Systematic biology is the comparative study of living and fossil species. It encompasses taxonomy, identification of species, elucidation of relationships between species, development of nomenclature, and classificatory protocols.

47. H. Shands, interview with author, tape recording, Beltsville, Md., 21 March 1996.

48. Borris, interview.

49. Ibid.

50. Belson, interview.

51. J. Kloppenburg and M. Balick, "Property Rights and Genetic Resources: A Framework for Analysis," in *Medicinal Resources from the Forest: Biodiversity and Its Importance to Human Health*, ed. M. Balick, E. Elisabetsky, and S. Laird (New York: Columbia University Press, 1996), 174–81.

52. Ibid.; my italics.

53. Grifo, interview.

54. Pfizer's promotional material about their 1993 project was available at http://www.pfizer.com/science/openingnaturefrm.html; accessed 23 February 1996.

55. In the interests of protecting the identity of this interviewee, his or her comments must remain unattributed.

56. Ibid.

57. United Nations Development Fund, *Statements of the Regional Meetings of Indigenous Representatives on the Conservation and Protection of Indigenous Knowledge* (Suva, Fiji: United Nations Development Fund, 1995), 3.

58. E. Hoagland, interview with author, tape recording, Washington, D.C., 20 March 1996.

59. Kress, interview.

60. Ibid.

61. Sytematics deals with the classification and categorization of natural specimens; systematics collections are those collated for this purpose.

62. The remainder confirmed that they did allow commercial access to their collections but did not note from what date.

63. Lisa Famolare, interview with author, tape recording, Washington, D.C., 21 March 1996.

64. King, interview.

65. The process of screening is eventually consumptive of the material resource.

66. Cragg, interview, 19 March 1996.

67. J. Rosenthal, interview with author, tape recording, Bethesda, Md., 18 March 1996.

68. Kress, interview.

69. M. Liteplo, telephone interview with author, tape recording, 13 March 1996.

70. Belson, interview.

71. B. Neirmann, interview with author, tape recording, Rockville, Md., 20 March 1996.

72. Incyte Press Release, 22 November 1999, PRNewswire.

73. Belson, interview.

74. Reproductive organs, for example.

6. TAMING THE SLIPPERY BEAST: REGULATING TRADE IN BIO-INFORMATION

1. G. Cragg, interview with author, tape recording, Bethesda, Md., 8 August 1995.

2. C. Peters, interview with author, tape recording, New York, 3 April 1997.

3. S. King, interview with author, tape recording, San Francisco, 28 September 1995.

4. T. Mays, et al. " 'Triangular Privity': A Working Paradigm for the Equitable Sharing of Benefits from Biodiversity Research and Development," in *Global Genetic Resources: Access, Ownership and Intellectual Property Rights*, ed. E. Hoagland and Y. Rossman (Washington, D.C.: Association of Systematics Collections, 1997), 287.

5. D. Downes, interview with author, tape recording, Washngton, D.C., 9 August 1995.

6. M. Foucault, *Power/Knowledge: Selected Interviews and Other Writing, 1972–77* (New York: Pantheon Books, 1980), 96.

7. Ibid.

8. The organization and operation of this elite network and my experiences in accessing it are described in greater detail in B. C. Parry, "Hunting the Gene-Hunters: The Role of Hybrid Networks, Status and Chance in Securing and Structuring Interviews with Corporate Elites," *Environment and Planning* 30 (1998): 2147–62.

9. M. Strathern, "Potential Property: Intellectual Property Rights and Property in Persons," *Social Anthropology* 4, no. 1 (1996): 24.

10. M. Balick, interview with author, tape recording, New York, 1 August 1995.

11. Downes, interview.

12. D. De Furia, interview with author, tape recording, Princeton, N.J., 24 June 1996.

13. R. Borris, interview with author, tape recording, Rahway, N.J., 29 April 1996; my italics.

14. Famolare is referring here to the agreement that was negotiated between Merck and InBio that was described in detail in Reid et al., eds. *Biodiversity*

Prospecting: Using Genetic Resources for Sustainable Development (Washington, D.C.: World Resources Institute, 1993). Interestingly, this agreement, which was negotiated in 1991, is based heavily on the principles established in the NCI's 1988/1989 version of the "Letter of Intent."

15. L. Famolare, interview with author, tape recording, Washington, D.C., 21 March 1996.

16. K. ten Kate and S. Laird, *The Commercial Use of Biodiversity: Access to Genetic Resources and Benefit-Sharing* (London: Earthscan, 1999), 64.

17. S. Laird, "Contracts for Biodiversity Prospecting," in *Biodiversity Prospecting: Using Genetic Resources for Sustainable Development*, ed. W. Reid et al. (Washington, D.C.: World Resources Institute, 1993), 108.

18. Balick, interview.

19. For a fuller description, see J. Rosenthal, "Equitable Sharing of Biodiversity Benefits: Agreements on Genetic Resources," Paper given at the OECD International Conference on Biodiversity Incentive Measures, Cairns, Australia, 25–28 March 1996, 6.

20. J. Rosenthal, interview with author, tape recording, Bethesda, Md., 18 March 1996.

21. De Furia, interview.

22. King, interview.

23. Ibid. Also see S. King, "Establishing Reciprocity: Biodiversity, Conservation, and New Models for Cooperation Between Forest Dwelling Peoples and the Pharmaceutical Industry," in *Intellectual Property Rights for Indigenous Peoples: A Sourcebook*, ed. T. Greaves (Oklahoma City: The Society for Applied Anthropology, 1994), 71–82; and K. Moran, "Health: Indigenous Knowledge, Equitable Benefits," *Indigenous Knowledge Notes*, no. 15 (December 1999), World Bank, Washington, D.C.

24. This inevitably varies from state to state. Some nations, such as the Philippines, Columbia, Peru, Ecuador, Venezuela, Bolivia, and Australia, have taken an active role in determining the form that compensation should take; others have not. Unfortunately, a detailed comparison of the different forms of compensation that have been secured in different countries and of the degree of active involvement that various states have played in this process is beyond the scope of this work.

25. J. Rosenthal, et al. "Combining High Risk Science with Ambitious Social and Economic Goals," *Pharmaceutical Biology* 37 (1999): S13.

26. Ibid.; and ten Kate and Laird, *The Commercial Use of Biodiversity*, 70–71.

27. There are, of course, some exceptions to this general rule. Shaman Pharmaceuticals gave money for "strengthening cultural development" in Peru by funding the Usko–Ayar Amazonian School of Painting. As Shaman suggests, however, this is much more than an art school: "While learning to paint local scenes of the forest, students also discover the names and uses of rain forest plants by recording the myths and traditional plant knowledge of their elders in painted form." K. See Moran, "Bio-cultural Diversity Conservation Through the Healing Forest Conser-

vancy," in *Intellectual Property Rights for Indigenous Peoples: A Sourcebook*, ed. T. Greaves (Oklahoma City: The Society for Applied Anthropology, 1994), 106.

28. Laird, "Contracts for Biodiversity Prospecting," 109.

29. Rosenthal et al., "Combining High Risk Science," S13.

30. C. Cook, quoted in J. Eberlee, "Assessing the Benefits of Bio-Prospecting in Costa Rica," *IDRC Reports*, 21 January 2000, available online at http://network.idrc.ca/ev.php?URL_ID=5571&URL_DO=DO_TOPIC&reload=1057275620.

31. Ibid.

32. M. Guerin-McManus, L. Famolare, I. Bowles, S. Malone, R. Mittermeier, and A. Rosenfeld, "Bioprospecting in Practice: A Case Study of the Suriname ICBG Project and Benefit Sharing under the Convention on Biological Diversity," report submitted to the Secretariat of the Convention on Biological Diversity, Montreal (1998): 17.

33. King, interview.

34. Shaman Pharmaceuticals, "Annual Report of 1994," 27.

35. Reported in E. Royte, "The Shaman and the Scientist," *San Francisco Focus* (August 1993): 128

36. Borris, interview.

37. Farncesca Grifo, interview with author, tape recording, New York, 28 March 1996.

38. Ibid.

39. K. Duffy Mazan, interview with author, tape recording, Bethesda, Md., 9 August 1995.

40. See D. Posey and G. Dutfield, *Beyond Intellectual Property: Towards Traditional Resource Rights for Indigenous and Local Communities* (Ottawa: IDRC, 1996); M. Battiste, M. Henderson, and J. Youngblood, *Protecting Indigenous Knowledge and Heritage: A Global Challenge* (Saskatchewan: Purich Publishing Ltd, 2000); M. Langton, *Burning Questions: Emerging Environmental Issues for Indigenous Peoples in Northern Australia* (Darwin, Australia: Center for Indigenous Natural and Cultural Resource Management, 1998).

41. Cragg, interview, 19 March 1996.

42. Borris, interview.

43. See for example, ten Kate and Laird, *The Commercial Use of Biodiversity*, 69.

44. Rosenthal et al., "Combining High Risk Science," 17; my italics.

45. See for example, C. Reed, ed., *Computer Law* (London: Blackstone Press, 1996).

46. Reed, *Computer Law*, 6.

47. See K. Moran, "Biocultural Diversity Conservation," 105.

48. King, interview.

49. Ibid.

50. Ten Kate and Laird, *The Commercial Use of Biodiversity*, 62.

51. Downes, interview.

52. H. Shands, interview with author, tape recording, Beltsville, Md., 21 March 1996.

53. N. Belson, telephone interview with author, tape recording, 10 May 1996.

54. See A. Sittenfeld and A. Artuso, "A Framework for Bio-diversity Prospecting: The InBio Experience," *Arid Lands Newsletter*, no. 37 (Spring/Summer 1995): 8–11.

55. This "Material Transfer Agreement" is available at http://www.wipo.org/tk/en/databases/contracts/texts/html/ncimta.html.

56. The role of MTAs as a mechanisms for preventing unlicensed use of collected bio-information are discussed further in B. C. Parry, "Mechanisms for Equitable Benefit Sharing Amongst Shareholders in Biodiversity Research and Development," in *Biodiversity 2000*, ed. C. Kheng and E. Lee (Kuching: The Sarawak BioDiversity Institute, 2001), 99–114.

57. Duffy-Mazan, interview.

58. Balick, interview.

59. Grifo, interview.

60. Shands, interview.

61. King, interview.

62. Borris, interview.

63. King, interview.

64. The auction announcement is available at http://archive.dovebid.com/brochure/SHAM010502.pdf.

65. The question of who "owns" or has property rights in such material is discussed further in B. C. Parry, "Bodily Transactions: Regulating a New Space of Flows in "Bio-Information," in *Property in Question: Appropriation, Recognition and Value Transformation in the Global Economy*, ed. K. Verdery and C. Humphrey (Oxford: Berg Press, 2003), 31–59.

66. Borris, interview.

67. Balick, interview.

68. Rosenthal et al., "Combining High Risk Science," 18

69. Duffy-Mazan, interview; my italics.

70. Rosenthal et al., "Combining High Risk Science," 18.

71. De Furia, interview.

72. "Invention" is here defined as the experiment undertaken to test the efficacy of the compound against a particular ailment.

73. Cragg, interview, 19 March 1996.

74. De Furia, interview.

75. See the most recent version of the NCI's "Memorandum of Understanding," available at http://www.dtp.nci.nih.gov/branches/npb/mou.doc.

76. R. Stone, "NIH Biodiversity Grants Could Benefit Shamans," *Science* 262, no. 5140 (10 December 1993): 1635.

77. De Furia, interview.

78. Cragg, interview, 19 March 1996.

79. Ibid.

80. Ibid.

81. Rosenthal, interview.

82. King, interview.

7. BACK TO THE FUTURE

1. The "Bonn Guidelines on Access to Genetic Resources and Fair and Equitable Sharing of the Benefits Arising out of Their Utilization" of the Convention on Biological Diversity is available at http://www.biodiv.org/doc/meetings/cop/cop-06/official/cop-06-06-en.doc.

2. Associated Press Release Online, "Theaters to Warn on Movie Piracy," 21 August 2002.

3. The livestock passport scheme is designed to regulate the transfer of individual cattle within the EU. See A. Kirby, "Form Filing Infuriates Farmers," BBC News Special, 14 September 1999, available at http://news.bbc.co.uk/1/hi/special_report/1999/09/99/farming_in_crisis/442052.stm.

4. O. O'Neill, *Autonomy and Trust in Bioethics* (Cambridge: Cambridge University Press, 2002), 130.

5. Ibid.

6. J. Thompson, *Political Scandal: Power and Visibility in the Media Age* (Cambridge: Polity, 2000), 254.

7. K. Ten Kate and S. Laird, *The Commercial Use of Biodiversity: Access to Genetic Resources and Benefit-Sharing* (London: Earthscan, 1999), 2.

8. For a fascinating discussion of the way in which property rights have been extended to encompass all manner of phenomena, including many forms of cultural property, see M. Brown, *Who Owns Culture? Native Rights and the Future of the Public Domain* (Cambridge, Mass.: Harvard University Press, 2003).

9. The Food and Agriculture Orgaization of the United Nations has considered introducing a superfund into which industrial countries and seed companies would pay a small percentage of profits from the sale of folk varieties to compensate farmers in developing countries. For a further discussion, see Rural Advancement Foundation International, *Conserving Indigenous Knowledge: Integrating Two Systems of Innovation* (New York: United Nations Development Program, 1994), 33; and D. Soleri et al., "Gifts from the Creator: Intellectual Property Rights and Folk Crop Varieties," in *Intellectual Property Rights for Indigenous Peoples: A Sourcebook*, ed. T. Greaves (Oklahoma City: The Society for Applied Anthropology, 1994), 25.

APPENDIX. METHODOLOGY

1. B. Parry, "Hunting the Gene-Hunters: The Role of Hybrid Networks, Status, and Chance in Securing and Structuring Interviews with Corporate Elites," *Environment and Planning* 30 (1998): 2147–62.

BIBLIOGRAPHY

Anderson, I. "Oceans Plundered in the Name of Medicine." *New Scientist* 148, no. 2005 (1995): 5.

Angerhofer, C., and J. Pezzuto. "Application of Biotechnology in Drug Discovery and Evaluation." In *Biotechnology and Pharmacy*, ed. J. Pezzuto, M. Johnson, and H. Manase, 312–65. London: Chapman and Hall, 1993.

Aoki, K. "Intellectual Property and Sovereignty: Notes Towards a Cultural Geography of Authorship." *Stanford Law Review* 48 (1996): 1293–1355.

Associated Press Release Online. "Theaters to warn on Movie Piracy" 20 August 2002.

Bains, W. "Biotechnology for Pleasure and Profit." *Bio/Technology* 11 (November 1993): 1243–47.

Balick, M. "Ethnobotany and the Identification of Therapeutic Agents from the Rainforest." In *Proceedings of the Ciba Foundation Symposium: Bioactive Compounds from Plants*, ed. the Ciba Foundation, 22–39. Chichester: Wiley, 1990.

Balick, M., Elisabetsky, E. and Laird, S., eds. *Medicinal Resources from the Forest: Biodiversity and Its Importance to Human Health.* New York: Columbia University Press, 1996.

Barton, J. H. "The Biodiversity Convention and the Flow of Scientific Information." In *Global Genetic Resources: Access, Ownership, and Intellectual Property Rights*, ed. K. E. Hoagland and A. Y. Rossman, 51–56. Washington, D.C.: The Beltsville Symposia in Agricultural Research/The Association of Systematics Research Publishers, 1997.

Battiste, M., M. Henderson, and J. Youngblood. *Protecting Indigenous Knowledge and Heritage: A Global Challenge.* Saskatchewan: Purich Publishing Ltd, 2000.

Bedini, S. *The Pope's Elephant* Nashville, Tenn.: Sanders and Company, 1998.

Belk, R. *Collecting in a Consumer Society.* London: Routledge, 1995.

Bell, D. *The Coming of Post-Industrial Society: A Venture in Social Forecasting.* Harmondsworth: Penguin, 1974.

Bent, S., et al. *IPR in Biotechnology World-wide.* New York: Stockton Press, 1987.

Bentley, L., and B. Sherman. *Intellectual Property Law.* Oxford: Oxford University Press, 2001.

Berdy, J. "The Discovery of New Antibiotic Microbial Metabolites: Screening and Identification." *Progress In Microbiology* 27 (1989): 3–25.

Biotechnology Industry Organization. "Report on Biotech Revenues, 1992–2001." Available at http://www.bio.org/investor/signs/200210rva.asp.

————. "Report on Revenues of Selected Companies, 1980–2001." Available at http://www.bio.org/investor/signs/200210rv5.asp.

Blum, E. "Making Biodiversity Conservation Profitable: A Case Study of the Merck–InBio Agreement." *Environment* 35, no. 4. (May 1993): 16–29.

Boehm, K. *The British Patent System.* Vol. 1, *Administration.* Cambridge: Cambridge University Press, 1967.

Bohm, D., and D. Peat. *Science, Order, and Creativity.* London: Routledge, 1989.

Booth, W. "U.S. Drug Firm Signs Up to Farm Tropical Forests." *Washington Post,* 21 September 1991.

Boyd, M., and K. Paull. "Some Practical Considerations and Applications of the National Cancer Institute in Vitro Anticancer Drug Discovery Screen." *Drug Development Research* 34 (1995): 91–109.

Boyle, J. *Shamans, Software, and Spleens: Law and the Construction of the Information Society.* Cambridge, Mass.: Harvard University Press, 1996.

Brockway, L. *Science and Colonial Expansion: The Role of the British Royal Botanical Gardens.* New York: Academic Press, 1979

Brown, M. *Who Owns Culture? Native Rights and the Future of the Public Domain.* Cambridge, Mass.: Harvard University Press, 2003.

Browne, J. "Biogeography and Empire." In *Cultures of Natural History,* ed. N. Jardine, J. Secord, and E. Sparry, 305–22. Cambridge: Cambridge University Press, 1996.

Brush, S. "A Non-Market Approach to Protecting Biological Resources." In *Intellectual Property Rights for Indigenous Peoples: A Sourcebook,* ed. T. Greaves, 131–45. Oklahoma City: The Society for Applied Anthropology, 1994.

Caminhoa, Joaquim Monteiro. *Memoria sobre o modo de conservar as plantas, com suas formas e cores, ou dos hervarios em geral e particularmente em liquidos.* Rio De Janeiro, 1873.

Castells, M. *The Informational City.* Oxford: Blackwell, 1989.

Chaterjee, P., and M. Finger. *The Earth Brokers: Power, Politics, and World Development.* London: Routledge, 1994.

Clifford, J. *The Predicament of Culture: Twentieth-Century Ethnography, Literature, and Art.* New York: Harvard Press, 1988.

Coombe, R. "Authorial Cartographies: Mapping Proprietary Borders in a Less-Than-Brave New World." *Stanford Law Review* 48 (1996): 1357–66.

Cornish, W. R. C. *Intellectual Property: Patents, Copyright, Trade Marks, and Allied Rights.* London: Sweet and Maxwell, 1999.

Cotton, C. M. *Ethnobotany: Principles and Applications.* Chichester: Wiley, 1996.

Cox, P. "Ethnopharmacology and the Search for New Drugs." In *Proceedings of the Ciba Foundation Symposium: Bioactive Compounds from Plants,* ed. the Ciba Foundation, 40–55. Chichester: Wiley, 1990.

Cragg, G., and P. Boyd. "Drug Discovery and Development at the National Cancer Institute: The Role of Natural Products of Plant Origin." In *Medicinal Re-*

sources of the Tropical Forest: Biodiversity and Its Importance to Human Health, ed. M. Balick, E. Elisabetsky, and S. Laird, 101–36. New York: Columbia University Press, 1996.

Cragg, G., R. Boyd, M. Grever, and S. Schepartz. "Policies for International Collaboration and Compensation in Drug Discovery and Development at the United States National Cancer Institute: The NCI Letter of Collection." In Intellectual Property Rights for Indigenous Peoples: A Sourcebook, ed. T. Greaves, 83–99. Oklahoma City: The Society for Applied Anthropology, 1994.

Cragg, G., et. al. "Natural Product Drug Discovery and Development at the National Cancer Institute: Policies for International Collaboration and Compensation." In Monographs in Systematic Botany from the Missouri Botanical Garden, vol. 48, 221–31. St. Louis: Missouri Botanical Gardens Press, 1994.

Crooke, S. "Optimizing the Impact of Genomics on Drug Discovery and Development." Nature Biotechnology 16 (1998): S2.

Cunningham, A. Ethics, Ethnobotanical Research, and Biodiversity. Geneva: WWF International, Gland Publicatons, 1993.

Daly, H., and R. Goodland. "An Ecological Assessment of Deregulation of International Commerce Under GATT." Environmental Working Paper Draft Document prepared for The World Bank Environment Department. Geneva: The World Bank, 1993.

Daston, L., and K. Park. Wonders and the Order of Nature, 1150 -1750. New York: Zone Books, 1998.

Davis, W. One River: Science, Adventure, and Hallucinogens in the Amazon Basin. London: Simon and Schuster, 1996.

Desmond, R., and F. Hepper. A Century of Kew Plantsmen: A Celebration of the Kew Guild. Kew, Richmond: Kew Guild Publishers.

Dicken, P. Global Shift: Transforming the World Economy. 3d ed. London: Chapman, 1999.

Di Masi, J., et al. "Cost of Innovation in the Pharmaceutical Sector." Journal of Health Economics 10 (1991): 107–42.

Downes, D. "The Convention on Biological Diversity and the GATT." In The Use of Trade Measures in Select Multilateral Environmental Agreements, ed. R. Housman et al., 197–251. Nairobi, Kenya: UNEP Publishers, 1995.

Doyle, J. Altered Harvest: Agriculture, Genetics, and the Fate of the World's Food Supply. New York: Viking, 1985.

Doyle, M. W. Empires. Ithaca, N.Y.: Cornell University Press, 1986.

Drahos, P. A Philosophy of Intellectual Property. Aldershot: Dartmouth, 1996.

Drayton, R. Nature's Government: Science, Imperial Britain, and the "Improvement" of the World. New Haven, Conn.: Yale University Press, 2000.

Drucker, P. Post-Capitalist Society. Oxford: Butterworth-Heinnemen, 1993.

Dutton, W., ed. Information and Communication Technologies: Visions and Realities. Oxford: Oxford University Press, 1996.

Eberlee, J. "Assessing the Benefits of Bio-Prospecting in Costa Rica." *IDRC Reports*, 21 January 2000. Available at http://network.idrc.ca/ev.php?URL_ID=5571&URL_DO=DO_TOPIC&reload =1057275620.

The Economist. "Biotech's Secret Garden." 30 May 1998, 75.

———. "From Blunderbuss to Magic Bullet." 21 February 1998, S9.

———. "Horn of Plenty: New Genetic Knowledge Means More and More Effective Drugs." *The Economist*, 21 February 1998, S5.

Eisner, T. "Prospecting for Nature's Chemical Riches." *Issues in Science and Technology* 6, no. 2 (Winter 1989): 31–34.

Ernste, H., and C. Jaeger. *Information Society and Spatial Structure.* London: Belhaven Press, 1989.

Farnsworth, N. "The Role of Ethnopharmacology in Drug Development." In *Proceedings of the Ciba Foundation Symposium: Bioactive Compounds from Plants*, ed. the Ciba Foundation, 2–21. Chichester: Wiley, 1990.

———. "Screening Plants for New Medicines." In *Biodiversity*, ed. E. O. Wilson, 83–97. Washington, D.C.: National Academy Press, 1988.

Featherstone, M. "Global Culture: An Introduction." In *Global Culture: Nationalism, Globalisation, and Modernity*, ed. M. Featherstone, 1–15. London: Sage Publications, 1990.

Findlen, P. "Courting Nature." In *Cultures of Natural History*, ed. N. Jardine, J. Secord, and E. Sparry, 57–75. Cambridge: Cambridge University Press, 1996.

———. *Possessing Nature: Museums, Collecting, and Scientific Culture in Early Modern Italy.* Berkeley: University of California Press, 1994.

Ford, R. I. "Ethnobotany: Historical Diversity and Synthesis." In *The Nature and Status of Ethnobotany*, ed. R. I. Ford, 33–49. Ann Arbor, Mich.: Museum of Anthropology Publication, 1978.

Forgan, S. "Context, Image, and Function: A Preliminary Enquiry into the Architecture of Scientific Societies." *The British Journal for the History of Science* 19 (1986): 89–113.

Foucault, M. *The Order of Things: An Archaeology of the Human Sciences.* New York: Vintage, 1970.

———. *Power/Knowledge: Selected Interviews and Other Writing, 1972–77.* New York: Pantheon Books, 1980.

Fox Keller, E. *The Century of the Gene.* Cambridge, Mass.: Harvard University Press, 2000.

Fuller, S. "Studying the Proprietary Grounds of Knowledge." *Journal of Social Behaviour and Personality* 6, no. 6, special issue, "To Have Possessions: A Handbook on Ownership and Property," ed. F. W. Rudmin (1991): 105–28.

Gamez, R. "A New Lease on Life." In *Biodiversity Prospecting: Using Genetic Resources for Sustainable Development*, ed. W. Reid et al., 1–52. Washington, D.C.: World Resources Institute, 1993.

Gasgoine, J. "The Ordering of Nature and the Ordering of Empire: A Commentary." In *Visions of Empire: Voyages, Botany, and Representations of Empire*, ed. D. Miller and P. Reill, 107–17 Cambridge: Cambridge University Press, 1996.

Gershon, D. "If Biodiversity Has a Price, Who Sets It and Who Should Benefit?" *Nature* 279 (1992): 284–95.

Glaxo Wellcome. "More and Better Candidate Medicines from Glaxo Wellcome R&D." 9 December 1997. Glaxo Press Release Archive, London. Available at http://www.gsk.com/press_archive/gw/1997/press_971209pr.htm.

Glowka, L. "A Guide to the Convention on Biological Diversity." New York: UNEP, 1993.

Goodman, D., and M. Redclift. *Refashioning Nature: Food, Ecology, and Nature*. London: Routledge, 1991.

Goodman, D., and J. Wilkinson. "Patterns of Research and Innovation in the Modern Agri-Food Sector." In *Technological Change and the Rural Environment*, ed. P. Lowe, T. Marsden, and S. Whatmore, 127–46. London: David Fulton Publishers, 1990.

Goodman, J., and V. Walsh. *The Story of Taxol: Nature and Politics in the Pursuit of an Anti-Cancer Drug*. Cambridge: Cambridge University Press, 2001.

Gordon, W. "A Property Right in Self-Expression: Equality and Individualism in the Natural Law of Intellectual Property." *Yale Law Journal* 102 (1993): 1533–57.

Greene, S. "Intellectual Property, Resources, or Territory? Reframing the Debate over Indigenous Rights, Traditional Knowledge, and Pharmaceutical Bioprospection." In *Truth Claims: Representation and Human Rights*, ed. Mark Philip Bradley and Patrice Petro, 229–49. New Brunswick, N.J.: Rutgers University Press, 2002.

Grever, M., S. Schepartz, and B. Chabner. "The National Cancer Institute: Cancer Drug Discovery and Development Program." *Seminars in Oncology* 19, no. 6 (1992): 622–38.

Grifo, F. T. "Chemical Prospecting: An Overview of the International Biodiversity Co-operative Groups Program." In *Emerging Connections: Biodiversity, Biotechnology, and Sustainable Development in Health and Agriculture*, ed. J. Feinsilver, 12–26. Washington, D.C.: Pan-American Health Organization, 1994.

Guerin-McManus, M., L. Famolare, I. Bowles, S. Malone, R. Mittermeier, and A. Rosenfeld. "Bioprospecting in Practice: A Case Study of the Suriname ICBG Project and Benefit Sharing Under the Convention on Biological Diversity." Report submitted to the Secretariat of the Convention on Biological Diversity, Montreal (1998): 1–24.

Haraway, D. *Modest_Witness@Second_Millennium.FemaleMan©_Meets_Onco-Mouse™: Feminism and Technoscience*. New York: Routledge, 1997.

——. *Symians, Cyborgs, and Women: The Re-invention of Nature*. New York: Routledge, 1991.

Hartl, D., and W. Jones. *Essential Genetics*. Boston: Jones and Bartlett Publishers, 1999.

Harvey, D. *The Condition Of Postmodernity: An Enquiry into the Origins of Cultural Change*. London: Basil Blackwell, 1990.

———. *Justice, Nature, and the Geography of Difference*. London: Basil Blackwell, 1996.

Hayden, C. "From Market to Market: Bioprospecting's Idioms of Inclusion." *American Ethnologist* 30, no. 1 (2003): 359–71.

———. *When Nature Goes Public: The Making and Unmaking of Bio-Prospecting in Mexico*. Princeton, N.J.: Princeton University Press, 2003.

Hildyard, N. "Foxes in Charge of the Chickens." In *Global Ecology: A New Area of Political Conflict*, ed. W. Sachs, 22–35. London: Zed Books, 1995.

Hix, J. *The Glasshouse*. New York: Phaidon, 1996.

Johnson-Laird, A. "Exploring the Information Superhighway: The Good, The Bad, and the Ugly." Paper presented at the Electronic Information Law Institute Meeting, San Francisco, 2–3 March 1995.

Joyce, C. "Prospectors for Tropical Medicines." *New Scientist*, 19 October 1991, 36–40.

———. "Western Medicine Men Return to the Field." *Bioscience* 42, no. 6 (1992): 399–403.

Juma, C. *The Gene Hunters* . London: Zed Books, 1989.

Katz, C. "Whose Nature, Whose Culture? Private Productions of Space and the 'Preservation' of Nature." In *Remaking Reality: Nature at the Millennium*, ed. Braun and N. Castree, 46–63. London: Routledge.

Kay, L. E. *Who Wrote the Book of Life: A History of the Genetic Code*. Stanford, Calif.: Stanford University Press, 2000.

———. "Who Wrote the Book of Life? Information and the Transformation of Molecular Biology, 1945–55." *Science in Context* 8, no. 4 (1995): 609–34.

Keeney, E. B. *The Botanists: Amateur Scientists in Nineteenth-Century America*. Chapel Hill: University of North Carolina Press, 1992.

King, S. "Antifungal in the Jungle." *Worth*, Winter 1994/1995, 87.

———. "Establishing Reciprocity: Biodiversity, Conservation, and New Models for Cooperation Between Forest Dwelling Peoples and the Pharmaceutical Industry." In *Intellectual Property Rights for Indigenous Peoples: A Sourcebook*, ed. T. Greaves, 71–82. Oklahoma City: The Society for Applied Anthropology, 1994.

Kirby, A. "Form Filing Infuriates Farmers." BBC News Special, 14 September 1999. Available at http://news.bbc.co.uk/1/hi/special_report/1999/09/99/farming_in_crisis/442052.stm.

Kloppenburg, J. *First the Seed: The Political Economy of Plant Biotechnology, 1492–2000*. Cambridge: Cambridge University Press, 1988.

Kloppenburg, J., and M. Balick. "Property Rights and Genetic Resources: A Framework for Analysis." In *Medicinal Resources from the Forest: Biodiversity and Its Importance to Human Health*, ed. M. Balick, E. Elisabetsky, and S. Laird, 174–81. New York: Columbia University Press, 1996.

Koerner, L. "Carl Linnaeus in His Time and Place." In *Cultures of Natural History*, ed. N. Jardine, J. Secord, and E. Sparry, 145–63. Cambridge: Cambridge University Press, 1996.

Laird, S. "Contracts for Biodiversity Prospecting." In *Biodiversity Prospecting: Using Genetic Resources for Sustainable Development*, ed. Reid et al., 99–130. Washington, D.C.: World Resources Institute, 1993.

Langton, M. *Burning Questions: Emerging Environmental Issues for Indigenous Peoples in Northern Australia*. Darwin, Australia: Center for Indigenous Natural and Cultural Resource Management, 1998.

Latour, B. *Science in Action: How to Follow Scientists and Engineers Through Society*. Cambridge, Mass.: Harvard University Press, 1987.

———. *We Have Never Been Modern*. London: Harvester Wheatsheaf, 1993.

Lemmon, K. *Golden Age of Plant Hunters*. London: Phoenix House, 1968.

Leutwyler, K. "Send in the Clones." *Scientific American*, 27 July 1998. Available at http://www.sciam.com/article.cfm?articleID=000186A6-697C-1CE2-95FB809EC588EF21.

Lewis, D. "The Gene Hunters." *The Geographical Magazine* 63, no. 1 (January 1991): 36–43.

Locke, J. *Two Treatises of Government*. Ed. P. Laslett. 1690. Reprint, Cambridge: Cambridge University Press, 1988.

Lyon, D. *The Information Society: Issues and Illusions*. Oxford: Polity Press, 1988.

Mackay, D. "Agents of Empire: The Banksian Collectors and Evaluation of New Lands." In *Visions of Empire: Voyages, Botany, and Representations of Empire*, ed. D. Miller and P. Reill, 38–58. Cambridge: Cambridge University Press, 1996.

MacLeod, D. "Survival of the Fittest." *The Guardian*, 19 May 1998.

Markus, T. *Buildings and Power: Freedom and Control in the Origins of Modern Building Types*. New York: New Brunswick, 1993.

Marsden, T., and S. Whatmore, eds. *Technological Change and the Rural Environment*. London: David Fulton Publishers, 1990.

Martin, E. "The End of the Body." *American Ethnologist* 19, no. 1 (1992): 121–40.

Massey, D., and J. Allen. *Geography Matters!* Cambridge: Cambridge University Press, 1984.

Mays, T., et al. " 'Triangular Privity': A Working Paradigm for the Equitable Sharing of Benefits from Biodiversity Research and Development." In *Global Genetic Resources: Access, Ownership, and Intellectual Property Rights*, ed. E. Hoagland and Y. Rossman, 279–98. Washington, D.C.: Association of Systematics Collections, 1997.

McChesney, J. "Biological Diversity, Chemical Diversity, and the Search for New Pharmaceuticals." Paper presented at the Symposium on Tropical Forest

Medical Resources and the Conservation of Biodiversity, Rainforest Alliance, New York, 1992.

McCloud, T., et al. "Extraction of Bioactive Molecules from Marine Organisms." Paper presented at The International Congress on Natural Products Research, Park City, Utah, 17–21 July 1988.

———. "Extraction of Bioactive Molecules from Plants." Paper presented at The International Congress on Natural Products Research, Park City, Utah, 17–21 July 1988.

McGraw-Hill Higher Education. "More Accessible Plant Alkaloids Through Cell Culture Systems." February 2000. Available at http://www.mhhe.com/biosci/pae/botany/botany_map/articles/article_05.html.

Middleton, N., P. O'Keefe, and S. Moyo. *Tears of the Crocodile: From Rio to Reality in the Developing World*. London: Pluto Press, 1993.

Miller, D. "Joseph Banks, Empire, and the 'Centres of Calculation' in Late Hanoverian London." In *Visions of Empire: Voyages, Botany, and Representations of Empire*, ed. D. Miller and P. Reill, 21–28. Cambridge: Cambridge University Press, 1996.

Mooney, P. R. "Exploiting Local Knowledge: International Policy Implications." In *Cultivating Knowledge: Genetic Diversity, Farmer Experimentation, and Crop Research*, ed. W. De Boef et al., 171–98. London: Intermediate Technology Publications, 1993.

———. "The Law of the Seed: Another Development in Plant Genetic Resources." *Development Dialogue*, nos. 1–2 (1983): 1–72.

———. *Seeds of the Earth: A Private or Public Resource?* Ottawa: Inter Pares Publishing, 1979.

Moran, K. "Bio-cultural Diversity Conservation Through the Healing Forest Conservancy." In *Intellectual Property Rights for Indigenous Peoples: A Sourcebook*, ed. T. Greaves, 101–9. Oklahoma City: The Society for Applied Anthropology, 1994.

———. "Health: Indigenous Knowledge, Equitable Benefits." *Indigenous Knowledge Notes*, no. 15 (December 1999). World Bank, Washington, D.C.

Mossinghoff, G., and T. Bombelles. "The Importance of Intellectual Property Protection to the American Research-Intensive Pharmaceutical Industry." *The Columbia Journal of World Business* 31, no. 1 (Spring 1996): 39–48.

MUD History. "Biotech Starts 2000 With A Bang as Recent Gains Rival Internet Stocks." http://www.bio.org/newsroom/newsitem.asp?id=2000_0120_01.

Mullaney, S. "Strange Things, Gross Terms, Curious Customs: The Rehearsal of Cultures in the Late Renaissance." *Representations* 3 (Summer 1983): 40–67.

Murdoch, J. "Inhuman/Nonhuman/Human: Actor-Network Theory and the Prospects for a Nondualistic and Symmetrical Perspective on Nature and Society." *Environment and Planning* 15, no. 6 (1997): 731–56.

Musson, A. E., and E. Robertson. *Science and Technology in the Industrial Revolution.* Manchester: Manchester University Press, 1969.

National Institutes of Health, National Institute of Mental Health, National Science Foundation, and the U.S. Agency for International Development. "International Cooperative Biodiversity Groups Agreement." 12 June 1992. Washington D.C.

The National Museum of Natural History, Smithsonian Institution. *Annual Report.* Washington, D.C., 1995.

Nature. "Names for Hi-jacking." Editorial, *Nature* 373, no. 6513 (February 1995): 370.

Newman, E. "Earth's Vanishing Medicine Cabinet: Rain Forest Destruction and Its Impact on the Pharmaceutical Industry." *American Journal of Law and Medicine* 20, no. 4 (1994): 479–91.

O'Neill, O. *Autonomy and Trust in Bioethics.* Cambridge: Cambridge University Press, 2002.

Outram, D. "New Spaces in Natural History." In *Cultures of Natural History*, ed. N. Jardine, J. Secord, and E. Spary, 249–65. Cambridge: Cambridge University Press, 1996.

Overbeck, W. *Major Principles of Media Law.* Philadelphia: Harcourt Brace College Publishers, 1996.

Parry, B. C. "Bodily Transactions: Regulating a New Space of Flows in 'Bio-Information.' " In *Property in Question: Appropriation, Recognition, and Value Transformation in the Global Economy*, ed. K. Verdery and C. Humphrey, 31–59. Oxford: Berg Press, 2003.

———. "The Fate of the Collections: Social Justice and the Annexation of Plant Genetic Resources." In *People, Plants, and Justice: The Politics of Nature Conservation*, ed. C. Zerner, 374–402. New York: Columbia University Press, 2000.

———. "Hunting the Gene-Hunters: The Role of Hybrid Networks, Status, and Chance in Securing and Structuring Interviews with Corporate Elites." *Environment and Planning* 30 (1998): 2147–62.

———. "Mechanisms for Equitable Benefit Sharing Amongst Shareholders in Biodiversity Research and Development." In *Biodiversity 2000*, ed. C. Kheng and E. Lee, 99–114. Kuching: The Sarawak BioDiversity Institute, 2001.

Pearce, S. *On Collecting: An Investigation Into Collecting in the European Tradition.* London: Routledge, 1995.

Pember, D. *Mass Media Law.* Chicago: Brown and Benchmark Publishers, 1997.

Phyton, Inc. "Phyton Expands Commercial Partnership with Bristol Meyers Squibb for Paclitaxel Supply." 1 July 2002. Available at: http://www.phyton-inc.com/frame_news7-10-02.html.

Pomian, K. *Collectors and Curiosities: Paris and Venice, 1500–1800.* Cambridge: Polity, 1990.

Posey, D., and G. Dutfield. *Beyond Intellectual Property: Towards Traditional Resource Rights for Indigenous and Local Communities*. Ottawa: IDRC, 1996.

Posey, D. A. "Intellectual Property Rights and Just Compensation for Indigenous People." *Anthropology Today* 6, no. 4 (1990): 13–16.

Power, P. "Interaction Between Biotechnology and the Patent System." *The Australian Intellectual Property Law Journal* 3, no. 4 (1992): 214–48.

Pratt, M-L. *Imperial Eyes: Travel Writing and Transculturation*. London: Routledge, 1992.

Price, C. "It Could Be Time to Face the Music: MP3 and Illicit Recordings." *The Financial Times*, 5 July 2000.

Radford, T. "The Breakthrough That Changes Everything." *The Guardian*, 26 June 2000.

Reed, C., ed. *Computer Law*. London: Blackstone Press, 1996.

Reid, W., et al., eds. *Biodiversity Prospecting: Using Genetic Resources for Sustainable Development*. Washington, D.C.: World Resources Institute, 1993.

Reilly, P. "Press Release from Conservation International: Forest People Search for New Medicines: Initiatives Builds Conservation-Based Industry." 7 December 1993, Washington, D.C.

Roberts, L. "Chemical Prospecting: Hope for Vanishing Ecosystems?" *Science* 256, no. 5060 (May 1992): 1142–43.

Robins, K., and D. Morley. *Spaces of Identity: Global Media, Electronic Landscapes, and Cultural Boundaries*. London: Routledge, 1995.

Rosenthal, J. "Equitable Sharing of Biodiversity Benefits: Agreements on Genetic Resources." Paper prepared for the OECD International Conference on Biodiversity Incentive Measures, 25–28 March 1996, Cairns, Australia.

Rosenthal, J., et al. "Combining High Risk Science with Ambitious Social and Economic Goals." *Pharmaceutical Biology* 37 (1999): S6–S21.

Royte, E. "The Shaman and the Scientist." *San Francisco Focus* (August 1993): 54, 128–29.

Rural Advancement Foundation International. "Bioprospecting and Indigenous Peoples: An Overview." Paper prepared for the South Pacific Consultation on Indigenous People's Knowledge and Intellectual Property Rights, United Nations Development Program, Fiji, 24–27 April 1995.

——. *Conserving Indigenous Knowledge: Integrating Two Systems of Innovation*. New York: United Nations Development Program, 1994.

Said, E. *Culture and Imperialism*. London: Vintage, 1993.

Schiebinger, L. "Gender and Natural History." In *Cultures of Natural History*, ed. N. Jardine, J. Secord, and E. Sparry, 163–78. Cambridge: Cambridge University Press, 1996.

Schierer, E., J. Bernoff, J. Sorley, and M. Gerson. *Content Out of Control: A Forrester TechStrategy Report*. Cambridge Mass.: Forester Publications, 2000.

Schoenberger, E. "The Corporate Interview as a Research Method in Economic Geography." *The Professional Geographer* 43 (1991): 180–90.

Serres, M. *Statues*. Paris: François Bourin, 1987.

Shaman Pharmaceuticals. "Annual Report of 1994."

Sherman, B., and L. Bentley. *Intellectual Property Law*. Oxford: Oxford University Press, 2001.

Shiva, V. *Monocultures of the Mind: Perspectives on Biodiversity and Biotechnology*. Penang, Malaysia: Third World Network Publishers, 1993.

———. "Why We Should Say 'No' to GATT TRIPs." *Third World Resurgence*, no. 39 (November 1993): 32–39.

Simon, E. "Innovation and Intellectual Property Protection: The Software Industry Perspective." *The Columbia Journal of World Business* 31, no. 1 (1996): 31–37.

Sittenfeld, A., and A. Artuso. "A Framework for Bio-diversity Prospecting: The InBio Experience." *Arid Lands Newsletter*, no. 37 (Spring/Summer 1995): 8–11.

Sittenfield, A., and A. Lovejoy. "Biodiversity Prospecting" *Our Planet* 6, no. 4 (1994): 20–21.

Sittenfield, A., and R. Villers. "Costa Rica's InBIO: Collaborative Biodiversity Research Agreements with the Pharmaceutical Industry." In *Principles of Conservation Management*, ed. G. K. Meffe and C. R. Carroll, 500–504. Sunderland, Mass.: Sinaver, 1993.

Smith, Sir J. E. *A Selection of the Correspondence of Linnaeus, and Other Naturalists, From the Original Manuscripts*. London: Longman, Hurst, Rees, Orme, and Brown, 1821.

Soleri, D., et al. "Gifts from the Creator: Intellectual Property Rights and Folk Crop Varieties." In *Intellectual Property Rights for Indigenous Peoples: A Sourcebook*, ed. T. Greaves, 21–37. Oklahoma City: The Society for Applied Anthropology, 1994.

Sterrckx, S. "Some Ethically Problematic Aspects of the Proposal for a Directive on the Legal Protection of Biotechnological Inventions." *European Intellectual Property Review* 123 (1998): 124–25.

Stevens, P. F. *The Development of Biological Systematics: Antoine-Laurent de Jussieu, Nature, and the Natural System*. New York: Columbia University Press, 1994.

Stevens, W. K. "Costa Rica in Pact to Search for Forest Drugs." *New York Times*, 24 September 1991.

Stone, R. "NIH Biodiversity Grants Could Benefit Shamans." *Science* 262, no. 5140 (10 December 1993): 1635.

Strathern, M. "Potential Property: Intellectual Property Rights and Property in Persons." *Social Anthropology* 4, no. 1 (1996): 17–32.

R. Sykes. "Research: A Key Factor for the Development of Countries." 1 December 1999. Available at http://www.fcs.es/fcs/eng/eidon/Introing/Eidon3/documentos/documentos.html.

Tangley, L. "Cataloguing Costa Rica's Diversity." *Bioscience* 40, no. 9 (1990): 633–36.

Ten Kate, K., and S. Laird. *The Commercial Use of Biodiversity: Access to Genetic Resources and Benefit-Sharing.* London: Earthscan, 1999.

Thomas, K. *Man and the Natural World: Changing Attitudes in England, 1500–1800.* London: Penguin, 1972.

Thompson, J. *Political Scandal: Power and Visibility in the Media Age.* Cambridge: Polity, 2000

Thompson, J. B. *Studies in the Theory of Ideology.* Cambridge: Polity Press, 1984.

Timmermann, B., and B. Huchinson. "Biodiversity Prospecting in the Drylands of Chile, Argentina, and Mexico." *Arid Lands Newsletter*, no. 37 (Spring/Summer 1995): 12–21.

Touraine, A. *The Postindustrial Society.* London: Wildwood House, 1974.

United Nations Development Fund. *Statements of the Regional Meetings of Indigenous Representatives on the Conservation and Protection of Indigenous Knowledge.* Suva, Fiji: United Nations Development Fund, 1995.

Venkat, C. "Paclitaxel Production Through Plant Cell Culture: An Exciting Approach to Harnessing Biodiversity." Paper presented at the International Conference on Biodiversity and Bioresources: Conservation and Utilization, Phuket, Thailand, 23–27 November 1997. Available at http://www.iupac.org/symposia/proceedings/phuket97/venkat.html.

Waldby, C. *The Visible Human Project: Informatic Bodies and Posthuman Medicine.* London: Routledge, 2000.

Wani, M. C., et al. "Plant Antitumor Agents. Part 6: The Isolation and Structure of Taxol, a Novel Antileukemic and Antitumor Agent from *Taxus brevifolia.*" *Journal of the American Chemistry Society* 93, no. 9 (1971): 2325–27.

Whitaker, K. "The Culture of Curiosity." In *Cultures of Natural History*, ed. N. Jardine, J. Secord, and E. Spary, 75–90. Cambridge: Cambridge University Press, 1996.

Wilkes, G. "The World's Crop Plant Germplasm: An Endangered Resource." *Bulletin of Atomic Scientists* 33 (1977): 8–16.

Woods, M., and A. Warren. *Glasshouses: A History of Greenhouses, Orangeries, and Conservatories.* New York: Rizzoli, 1988.

World Commission on Environment and Development. *Our Common Future.* Oxford: Oxford University Press, 1987.

INDEX

academic and scientific institutions, 105, 127, 130, 233; corporate access to collections, 183–88, 234; funding cuts, 129–30, 181, 187. *See also* New York Botanical Garden; Smithsonian
access, 60–61, 176–77, 181–88, 234, 252
accountability, 257–58
accumulation, cycles of, 19–22, 30
acquisition, 15, 16–19
advance payments, 118–19, 206, 214–17, 247
Agenda 21, 94
agents of empire, 29, 36–37
Agreement Establishing the World Trade Organization, The, 79
Agriculture Department (USDA), 38–39, 105–6, 175, 236
Aldorvandi, Ulisse, 20
American Museum of Natural History, 105, 174, 236
American Type Culture Collection (ATCC), 139, 195
Angerhofer, C., 114
animal models, 114–15
Aristotle, 19
Arnold Arboretum, 107
article 35, section 101 (U.S. Constitution), 86
artifactual forms, 5, 7, 9, 23, 58–59, 103, 200; dematerialization, 59, 65–66
artificialia, 18
Association of Systematics Collections (ASC), 180, 183, 268
AstraZeneca, 196
authorship, 225–26

Balick, Michael, 124, 134, 178, 203, 210–12, 215
Banks, Joseph, 33–34
Belize, 134
Bell, Daniel, 50, 54
Belson, Neil, 166–67, 173, 194–95, 197, 232, 234

benefit-sharing agreements. *See* compensation
Bentley, L., 83, 88
biocolonialism, 117, 118, 123, 251
Biodiversity Convention. *See* Convention on Biological Diversity
Biodiversity Inventory, 133
bio-imperialism, 13–14, 93–94
bio-informatics, 162
bio-information, xx, 250; combinations and permutations, 160–66; as commodity, 162–63; licensing and "pay-per-view," 195–99; proxies, 62–69; regulation of trade, 78–101; transmissibility, 100–101; unlicensed copying of, 197, 233–39; untraceable trade in, 151, 182. *See also* biological materials
biological materials: as alienable commodities, 102, 135, 148, 155, 200; combination of, 72–73; constituent parts, 47–49, 64; decorporealization, 59, 65–66, 72–74, 115, 190, 195–98, 200–201, 228; disciplining of, 29–31, 92, 240; informational forms, 68–69, 74–75; information-based metaphors, 51–53, 63–68; interior worlds of, 47–49, 64, 135; nineteenth-century valuation of, 45–53; patrimony of nation-states, 118, 155; space of flows in, 30–32, 275nn.58, 62; terminology, 5, 51–53, 65, 68, 96. *See also* bio-information; proxies; raw materials
bioprospecting, 2, 4; 1985–1995, 131–35; consortia, 126–29, 192–93, 286n.58; environmental rationale, 94–95, 120–22, 126–27, 129, 169–70; ethnobotany, 133–34; imagery of colonial era, 12–13; intensification of, 131, 134–35; linkages between neophytes and established collectors, 126, 131, 177–78, 200; networks, 109–10, 126, 128; stockpiling, 171–72; transformations in, 42–43, 147–48,